Technonatures

ENVIRONMENTAL
HUMANITIES

Technonatures

Environments, Technologies, Spaces, and Places in the Twenty-first Century

Damian F. White and Chris Wilbert, editors

Wilfrid Laurier University Press

[WLU]

We acknowledge the financial support of the Government of Canada through the Book Publishing Industry Development Program for our publishing activities.

Library and Archives Canada Cataloguing in Publication

Technonatures : environments, technologies, spaces, and places in the twenty-first century / edited by Damian F. White and Chris Wilbert.

(Environmental humanities series)
Includes bibliographical references and index.
ISBN 978-1-55458-150-4

1. Environmental sciences — Philosophy.   2. Nature — Effect of human beings on.   3. Environmentalism — Philosophy.   4. Political ecology — Philosophy.   I. White, Damian F., 1967–   II. Wilbert, Chris, [date]   III. Series: Environmental humanities series

GE40.T42 2009   333.701   C2008-907972-8

© 2009 Wilfrid Laurier University Press
Waterloo, Ontario, Canada
www.wlupress.wlu.ca

The cover photograph shows a sculpture by Rosario López Parra titled *Guatavita 04* (2007). The sculptor was born in Bogatá in 1970 and is a professor in the Fine Arts Department at the Universidad Nacional de Columbia. Cover design by David Drummond. Text design by Daiva Villa, Chris Rowat Design.

Contents

Preface

Environmentalism and the environmental social sciences appear to be in a period of disorientation and even transition. This collection draws together leading international thinkers to explore the notion that perhaps one explanation for the current malaise of the "politics of ecology" is that we increasingly find ourselves negotiating "technonatural" time/spaces.

Technonatures provides a definitive mapping of a series of new political ecologies that are unfolding in the environmental social sciences, the environmental humanities, and technology studies, leaving anxieties and consternation in their wake. Contributors variously argue that a "technonatural sensibility" can be found in attempts to grapple with the environmental consequences of emerging informational capitalism and network societies, as well as in growing concerns about the rise of biopolitical economies and the technological colonization of fleshy bodies. That same sensibility can be found in attempts to map the shifting relations between humans and non-humans and in ongoing debates about the future of humanism for environmental analysis. *Technonatures*, though, does not simply outline a narrative of disenchantment. Contributors also suggest that from the growing interest in ecotechnology, industrial ecology, and sustainable architecture to "hybrid aesthetics," from debates about "cosmopolitics" to recent attempts to recover the city as the ultimate technological–natural–social entity, multiple possibilities exist for constructing alternative technonatural futures.

Many of the themes addressed in this collection may be dismissed by traditional environmentalists as merely symptomatic of a colonizing technological idealism. Yet all the contributors to this collection argue in different ways that there has been a discernable shift in the imaginative horizons of the environmental debate of late. A central structure of feeling of "technonatural time/spaces" is that neither the environmental social sciences, nor the environmental humanities, nor environmentalism as a broad and diverse social movement, can understand or transform our social-ecological worlds in more egalitarian or "just" fashions by premising their narratives on a fictitious "organic" past or on nostalgic normative visions of purified "ecological" futures.

This book sketches out the contours and tensions of emerging technonatural political ecologies with the aid of contributions from a broad range of writers situated in a variety of disciplines and theoretical traditions. *Technonatures* brings together contributions, from authors variously influenced by Frankfurt School critical theory to those influenced by the sociologies of ecological modernization; from eco-Marxism to Deleuzian currents and actor network theory; from environmental justice, feminist, and anti-racist scholars to authors influenced by post-industrial and ecotechnological discourses, to demonstrate how a variety of perspectives are addressing technonatural time/spaces.

Acknowledgments

As is the case with many collective academic endeavours, this book emerged from a mixture of motives: the boredom that comes with a long summer spent reading exciting new materials but having few broader allies to engage with; a desire to think beyond the isolation and often narcissism of much academic life; and the hope of creating a context for thinking and acting collectively to build alliances that would allow us to more effectively act politically. The first manifestation of this urge arose back in 2001, at what was somewhat pretentiously called the "New Political Economy Reading Group," an interdisciplinary group of sociologists, geographers, and ecologists who met in the summer of 2001 at City University, London, out of a desire to engage with political economy, cultures, and natures. A number of themes dominated discussions at this group. Initially, there was a collective desire to read about and come to grips with a range of currents then surging though the environmental social sciences — currents inspired variously by the work of Henri Lefebvre, Bruno Latour, Doreen Massey, John Urry, Neil Smith, David Harvey, and Erik Swyngedouw. Soon the work of Gilles Deleuze, Donna Haraway, Andrew Barry, Sarah Whatmore, David Hess, and Bruno Latour, and developments in science and technology studies more generally, forced their way onto the agenda. Our practising ecologists suggested that we attend to the interesting work emerging from non-equilibrium ecologists; others suggested that social theory, political ecology, and STS were all very well but that someone should be keeping an eye on exciting developments occurring outside the social sciences, in industrial ecology and sustainable architecture. But the group settled most notably on a collective appreciation of the work of Noel Castree and Bruce Braun. Yet if one defining theme of this group was excitement over this innovative literature, which seemed to be allowing us to "think" the politics of nature in all kinds of promising new ways, there was equally a certain dismay at the extent to which this exciting intellectual culture of ferment over the politics of nature seemed to be entirely unrepresented in the increasingly conservative public discourse of environmentalism. As the energies of summer became overtaken by the demands of work and teaching, this forum fell apart; even so, the interest and concerns continued and were reworked and rethought in different contexts. And at one of the many conferences in the circuits of academe we decided to develop some events with the goal of working toward a different, more open, and more questioning politics of environmentalism.

Most notably, these thoughts were channelled across four "Technonatures" conferences that were held respectively at the Department of Sociology, Goldsmith's College, University of London in 2003; the Department of Geography, Oxford University, in 2004; the International Sociological Association Annual Meeting in Stockholm, in 2005; and the American Association of Geographers in Chicago, in 2006. In all four places the conversations continued. This book is largely the product of them.

As is often the case with such a long journey, this book represents many conversations held at many different times and is in debt to many different people. As such there are numerous people that need to be acknowledged. We would first like to thank members of the New Political Economy Reading Group for their thoughts and contributions in helping us develop the kernel of this project. Mike Michael made the first Technonatures conference possible through his generous support, and we would also like to thank Andrew Barry, Kate Soper, Sarah Whatmore, Steve Hinchliffe, Elizabeth Shrove, Tiziana Terranova, Andrew Jamison, and Caroline Bassett for taking part in that first symposium, along with the audience, who added immeasurably to the development of this project. Special thanks also go to Rikke Hansen and Lisbeth Tarri for their organizational help on the day.

Erik Swyngedouw co-organized the second conference with us at the Department of Geography at Oxford University. Generous funding for this was provided by the Social and Cultural Geography Research Group at the Royal Geographical Society/IBG as well as by James Madison University. We would like to thank Mary Lou Wiley, Doug Brown, Noel Castree, Alan Rudy, Fletcher Linder, Tim Forsyth, John Urry, Gail Davies, Bronwyn Parry, and (the virtual) Cathy Gere), as well as all the other participants, for making this such an enjoyable event. Thanks also go to Catherine Friel for great visuals.

Technonatures III in Stockholm was co-organized between us with the added help of Fletcher Linder. Technonatures IV in Chicago was co-organized with Erik Swyngedouw. There were too many participants at both these meetings for us to thank individually now, but a special thanks to Aidan Davison for insightful discussions in Stockholm down at the pub afterwards.

The first manifestation of the project emerged as a special issue of *Science as Culture* in 2006. We would like to thank the commissioning editor Les Levidow for his editing skills (and his seemingly endless recalcitrance, which taught us a thing or two), as well as the contributors to this book. We would additionally like to thank all the colleagues who reviewed papers and book chapters for the various Technonatures projects and all the contributors to this book for making this project possible.

On a personal level, Damian would like to thank Sarah Friel, Finbar White, Cormac White, Xavier White, Mary White, Mark White, Colin Butler, Nathalie Gordon and Mike Clark, Kay and Mark Davis, and all the Friels. For helping him think more clearly about technonatural matters, he would like to thank Ted Benton and the Red-Green Study Group, Neil Curry, Geoff Robinson, Alan Rudy, Nicole Merola, Jennifer Coffman, Aidan Davison, and all his friends and former colleagues at James Madison University, but especially Joe Rumbo, Nikita Imani, Joe Spear, Liam Buckley, John Ott, Sarah Warren, and the "Think" reading group: Fletcher Linder, David Ehrenpreis, Kevin Borg, and Steve Reich.

Chris would like to thank Steve Hinchliffe, Sarah Whatmore, Rikke Hansen, Caroline Bassett, Chris Philo, and the members of the Animal Studies Group for their support and intellectual stimulation. And he would like to remember the late Duncan Fuller, who was always a good person to be energized by about political spaces in the sometimes stultifying realms of academe.

Damian White
Providence, Rhode Island
dwhite01@risd.edu

Chris Wilbert
London
chris.wilbert@anglia.ac.uk

Introduction

Inhabiting Technonatural Time/Spaces
Damian F. White and Chris Wilbert

The expansion of mankind, both in numbers and per capita exploitation of Earth's resources, has been astounding. To give a few examples: During the past three centuries human population increased tenfold to 6000 million, accompanied e.g. by a growth in cattle population to 1400 million (about one cow per average size family). Urbanisation has even increased tenfold in the past century. In a few generations mankind is exhausting the fossil fuels that were generated over several hundred million years. The release of SO_2, globally about 160 Tg/year to the atmosphere by coal and oil burning, is at least two times larger than the sum of all natural emissions, occurring mainly as marine dimethyl-sulfide from the ocean... Considering these and many other major and still growing impacts of human activities on earth and atmosphere, and at all, including global, scales, it seems to us more than appropriate to emphasize the central role of mankind in geology and ecology by proposing to use the term "anthropocene" for the current geological epoch.
   —Crutzen and Stoermer (2000)

Increasingly in future... the time will come, for example, when massive programmes will have to be set in train to regulate the relationship between oxygen, ozone, and carbon dioxide in the earth's atmosphere. In this perspective, environmental ecology could equally be re-named "machinic ecology," since both cosmic and human practice are nothing if not machinic—indeed they are machines of war, in so far as "Nature" has always been at war with life!
   —Felix Guattari, *The Three Ecologies* (2000)

Political ecology, at least in its theories, has to let go of nature. Indeed, nature is the chief obstacle that has always hampered the development of public discourse.
— Bruno Latour, *The Politics of Nature* (2004)

... far from being dead and buried, nature is currently being practiced anew... But that nature is not what we have imagined it to be, fixed in its identity and unrelated to society.
— Steve Hinchliffe, *Geographies of Nature* (2008)

The Environmental Debate in Changing Times?

The current state of "the environmental debate" is in considerable flux at the beginning of the twenty-first century. From European Union countries to Argentina, from India to Canada, or China, Egypt, and beyond, diverse societies find themselves gripped by controversies, dilemmas, and disputes emerging from the incorporation and resistance of human and non-human bodies, ecologies and landscapes into circuits of commodification, property regulation, innovation, patenting, and enclosure. Disputes surrounding nanotechnology, biotechnology, and global warming, and concerns over biodiversity, water resources, and food—to name just a few acute issues—lend credence to the perception that natures, societies, and technologies are being jointly made and remade at dizzying speeds (Braun and Castree 1998). According to many writers (Haraway 1991, 1998; Luke 1997, 1999; Guattari 2000; Braun and Castree 1998), a seemingly unbounded, technologically instilled, and ideologically renewed capitalism appears to be intensifying the creative destruction of diverse ecologies around the globe even as it lurches unsteadily from boom to bust. Yet the movements that have been at the centre of politicizing these processes of remaking—notably the diverse ecological and green social movements that exploded onto the political scene with so much force in the last quarter of the twentieth century—seem politically and intellectually disorientated by such developments. Indeed, if we follow the thoughts of Bruno Latour, "the politics of nature" is increasingly marked by a degree of stagnation (2004, 1).

Latour is of course a leading provocateur whose work is defined by a penchant for the dramatic (see Castree 2006). Yet, at a time when the "environmental question" is at least rhetorically moving toward centre stage in the political world, the claim that there has concurrently been a loss of confidence, coherence, and vigour among certain manifestations of environmentalism is an assertion that has been reiterated recently by a much broader array of academic and activist voices, from different parts of the globe.

Some have pointed to the nervous and unsteady responses in Europe to the controversies raised by Bjorn Lomborg's *The Skeptical Environmentalist* (2001) and the subsequent Lomborg affair.[1] Others have suggested a certain plateauing of support can be detected in public opinion surveys for mainstream environmental organizations from the UK to Australia (MacNaghten 2003, 63; Davison in this volume). In the United States, it is Shellenberger and Nordhaus's internal critique of the mighty US environmental movement "The Death of Environmentalism" (2005) that has most crystallized concerns. While maintaining that mainstream environmental movements in the US have made important regulatory gains over the last three decades in the fight for basic environmental protection, Shellenberger and Nordhaus have suggested over more recent decades — and with particular reference to global warming — that little further progress has been made. The dominant US environmental groups, they contend, are failing to generate a credible vision of the future or the political alliances that could bring "progress" about. More recently, they have refined this critique (Nordhaus and Shellenberger 2007) to argue that the manner in which much conventional mainstream environmental critique has relied on a narrative which problematizes human agencies within the context of a static and a-historical image of "Nature" has lead to a "politics of limits" that itself has significantly constrained the imaginative capacities to rethink a productive, progressive politics of the environment.

The "death of environmentalism debate" in the US and talk of post-environmentalism more generally might be premature. Environmental politics has often proved much more internally varied, mutable and much more resilient than its contrarian or post-environmental critics have acknowledged. However, what seems undeniable is that all these developments reflect a palpable sense of dissatisfaction with a certain style of environmental critique — romantic, ecocentric, and often neo-Malthusian — that has arguably dominated the imaginative horizons of much mainstream "northern" environmentalism over the last three decades.[2] We might identify this style of critique as a world view that while placing much emphasis on the virtues of "holism" nevertheless all too quickly defaults to a view that the world can be sharply separated into the discrete and singular categories of Society and Nature.[3] This mode of critique is more often than not universalising in terms of its descriptions and prescriptions. It can take on a somewhat scolding tone in relation to the question of agency (human or otherwise) and is somewhat neo-Calvinist in its visions of what a sustainable future should look like. In short, perhaps what is in crisis, then, is not environmental politics in general in all their complexities and pluralities but an environmental politics that has been premised on a naturalistic politics of Nature. As Hinchliffe (2007, 188) has observed, it is attempts to develop a politics that evokes Nature with a capital "N" that "doesn't seem to be working as a rallying site for everyone and everything anymore." Yet, what would seem to be emerging from this dissatisfaction with mainstream environmentalism are new politics of the environment that are struggling to define, engage, and enact multiple natures.

Many more voices are now emerging to argue that if we wish to see a truly inclusive "global" environmental debate, we need a more complex conversation than one that merely echoes and globalizes the latest anxieties of the mainstream political classes of the US and the EU or merely seeks to univeralize what are often quite culturally specific ideas about the appropriate natures we (and the "we" really becomes a problem here, as it has been for some time, in that it reflects most often a developed world conception of global interests) should defend and conserve. Literature and practices in environmental justice and political ecology have led the way, arguing that what is needed are environmental debates and politics that more adequately capture the diverse social framings and socio-ecological concerns of blue-collar workers, women and minority communities, urban, suburban, and ex-urbanites in the North (Bullard 1990; Faber 1998; Gandy 2002; Gottlieb 2002; Agyeman 2005; Jones 2008), and the equally diverse socio-ecological concerns of peasants, farmers, women, and the urban poor of the South (Agarwal and Narain 1991; Peet and Watts 1996; Rocheleau, Thomas-Slayter, and Wangari 1996; Guha and Martinez-Allier 1997; Forsyth 2002; Biersack and Greenberg 2006).

If the rise of literatures and activism around environmental justice and political ecology can be seen as marking the first and second movements of a revisionist wave moving through environmental studies, this book suggests that we can see a third (complementary) movement in a converging set of overlapping conversations opening up among currents of the environmental social sciences, the emerging field of the environmental humanities, and technology studies. While these conversations are still fairly heterogeneous, they share a starting point — namely, a sense that the source of many contemporary Green anxieties is that they are based on an implicit, albeit reluctant, recognition that critical discourses formulated around strong, power-laden oppositions and distinctions (of the organic versus the synthetic, the human versus the "animal" or machine, the natural versus the technological, etc.) have become not just much harder to maintain in recent years but also less politically desirable to maintain. As such, this book seeks to name a common *structure of feeling* that is emerging in diverse places and spaces. That is, many people increasingly find themselves negotiating "technonatural forms of life" (cf. Lash 2001) — indeed, they have been doing so for some time — and this needs to be reflected in ongoing calls for reworked political-ecological practices and theories.

Technonatures?

But why "technonatures"? What does this term point to? Why, in our current unevenly globalized time/spaces, when technological determinism is still rife, when an increasingly geneticized biology is so often invoked to provide the last word on the "natural" facts of human variations (Franklin 2000), and when every few years new technologically oriented epochs seem to be evoked, why conjoin "techno" with "natures"?

The term "technonatures" is deliberatively provocative and has numerous lineages. It is clearly in debt to a long line of cultural theory and environmental history from Raymond Williams to William Cronon (Williams 1973; Cronon 1995; Soper 1995) that in emphasizing the central role that social power has played in the constitution of landscape and our environments more generally has long cast a skeptical eye over the idea that a politics of the environment can be usefully grounded in terms of the rhetoric of defending the pure, the authentic, or an idealized past — that is, solely in terms of the ecological or the natural. It is a term that is clearly engaging (if somewhat critically) with the various attempts to capture the anthropogenic reach of modern humanity, whether this is posed in terms of the "humanization of nature" that has occupied a central place in the writings of Ulrich Beck in sociology or Thomas Hughes's discussion of a "human built world" in the history of technology (Beck 1992; Hughes

2004). Additionally, it is a discussion that is deeply indebted to the work of geographers such as Neil Smith, Noel Castree, and Bruce Braun, who have so cogently mapped the extent to which human societies have been involved in "the production of nature," giving rise to what they refer to as "social natures" (Smith 1984; Castree 1995; Braun and Castree 1998).

In deploying the term "technonatures" as an organizing myth and metaphor for thinking about the politics of nature in contemporary times, what we seek to do in this collection is explore one increasingly pronounced dimension of the social-natures discussion. Here, the term "technonatures" seeks to highlight a growing range of voices ruminating over the claim not only that we are inhabiting diverse social natures but also that knowledges of our worlds are, within such social natures, ever more technologically mediated, produced, enacted, and contested, and, furthermore, that diverse peoples find themselves, or perceive themselves, as ever more *entangled* with things—that is, with technological, ecological, cultural, urban, and ecological networks and diverse hybrid materialities and non-human agencies (Haraway 1991; Latour 1993; Wark 1994; Swyngedouw 1996; Luke 1997, 1999; Escobar 1999; Philo and Wilbert 2000; Michael 2000; Barry 2001; Lash 2001; Gandy 2002, 2004, 2005; Milani 2002; Kull 2002; Kaïka 2004). This is a politics of nature(s) that, as Steve Hinchliffe has observed, not only seeks to capture how "natures" and "societies" are interwoven in "a variety of different ways with a variety of different effects" but that non-humans of all kinds are "active and lively partners" in the making of our worlds (Hinchliffe 2007, 1, but additionally see Philo and Wilbert 2000; Hobson 2007).

Moreover, many contributors to this book argue it is not just a growing range of philosophers, historians, geographers, sociologists, and natural scientists but a growing range of activist currents as well that are taking seriously Donna Haraway's (1991) claim that our sensuous, embodied engagements with the world are not simply marked by collisions or, better still, by intra-actions between the organic and the synthetic. In addition to that, there is no sense of nature, of human subjectivity, of the body (nor are there even concepts of sustainability and ecology) that can effectively be thought outside of, or separable from, ever more technologized societies and social relations.

Clearly, claims like this are broad and general, with something of a sensationalist inflection. As such, they invite a skeptical response. Specifically, one might wonder what exactly is strikingly new about the "sudden" discovery of technologies in the environmental debate or indeed of recognizing the role played by multiple human and non-human agencies in this world? What exactly is new about the idea that "we" live in hybrid worlds comprising natures, societies, and technologies? One indeed might wonder here whether we haven't always lived this way. If we turn to Leo Marx's exploration of the themes of nature, technology, and culture in American literature, *The Machine in the Garden* (1964), we find that he illustrates how at the dawn of mass industrialism in the United States we can find in writings from Hawthorne to Melville, from Twain to Cooper, concerns about an emerging technological world "ripping through" and "unbalancing" a supposed natural order established between society and nature. What, then, is *new* about current technonatural talk and praxis? This is a complex question, one that is answered in this collection in different ways.

Technonatural Political Economies/Political Ecologies

Reflections on the changing nature of capitalist political economies have provided one key source for technonatural concerns over the past decade. That capital has embarked on a decisive phase of global restructuring over the past three decades has become axiomatic in much of contemporary social theory. But how, specifically, might such developments affect the horizons of the environmental debate? More and more voices have been arguing that grasping material transformations at the level of political economy is of central importance for rethinking political ecology in the twenty-first century. Continuing this logic, it is argued that the environmental social sciences in particular need to anticipate the environmental consequences of global political economies, which are characterized by a growing emphasis on production based on information, communication, and biotechnologies in the affluent core (Haraway 1991; Castells 1996; Hardt and Negri 2000; Luke 1997; Thrift 2005) and characterized as well by the spatial and scalar repositioning of the principle centres of capitalist manufacturing (most obviously to China, most recently to India; Frank 1998), by increasingly rapid urbanization and suburbanization (Davis 1998, 2002), and by the reworking of cultural, political, and economic power into more liquid, rhizomatic, networked, or "fluid" forms (see Castells 1996; Bauman, 2000; Urry 2000, 2007; Hardt and Negri 2000; Thrift 2005).

For some, these transformed political economies are giving rise to a need for distinctly materialist technonatural political ecologies because our understanding of the social-ecological metabolism is now being significantly transformed. This is so, it is argued, because modern currents of environmental history increasingly suggest that even humanity's earliest ancestors were much greater shapers and transformers of the physical environment than environmentalist ideologies have previously allowed (Devevan 1992; Philips and Mighall 2000) and also because the "natures" now emerging from these contemporary processes of "creative destruction" are increasingly worked and reworked in the context of capitalist production.

Consider how capital is burrowing down into the micro-scales of bodies (human and non-human) through biotechnology and nanotechnology and burrowing up through industrial and agro-industrial transformations of historically unprecedented speed and scale (Braun and Castree 1998). As a result, according to one classic framing of the issue, human beings in modern times have found themselves, literally, constructing "a third nature." Applying the classic distinction between first (organic) nature and second (humanized) nature that runs through critical theory, MacKenzie Wark has vividly and imaginatively suggested that "second nature, which appears to us as the geography of cities and roads and harbours and wool stores is progressively overlaid with a third nature of information flows creating an information landscape which covers old territories" (Wark 1994, as cited in Braun and Castree 1998, 4).

Other perspectives on the political-economy debate posit that the environmental consequences of "informational capitalism" invite reflection on the potentialities and difficulties posed by state governance of environmental flows. Informational capitalism, it is argued, may accelerate tendencies toward demateralization (Mol 2003) or processes of environmental displacement (Harvey 1996). More generally, these currents suggest that attention must be paid to the specific problems of governance generated by a world that is marked by increasing flows (see Mol 2003; Oosterveer in this book). From this perspective, it is becoming increasingly important to develop a "sociology of flows" in order to understand the problems and dilemmas that beset efforts to govern the diverse environmental mobilities of people, animals, viruses, foods, wastes, goods, currencies, images, and symbols as they are funnelled through global networks — networks that are centred on the hubs of global or world cities and on the high-tech "nodes" or hinterlands that service those power centres.

There is no doubt that in much of the discussion about emerging "technonatural political economies" there is an epochal inflection (that can mimic archaeological notions of iron ages, bronze ages, etc.), notwithstanding disagreements about when such a rupture actually emerged. Wark (1993) suggests that the emergence of "third nature" can be traced back to the development of the telegraph. Haraway (1991) locates the rise of the "cyborg" in a post–Second World War historical moment characterized by the consolidation of interests in military technology, cybernetics, and space exploration. Castells (1996) traces the emergence of the "network society" to the rise of three independent processes: the information-technology revolution, the economic crisis of capitalism and statism, and the flourishing of new social movements from the mid-1960s to the early 1970s.

But there are many others who are weary of the rhetoric of epochal shifts and skeptical of such manoeuvres. Andrew Barry (2001), for example, has cautioned against technonatural epochal narratives. He maintains we do indeed find ourselves now in "technological societies." Moreover, "to analyse the conduct of political and economic life without considering the importance of material and immaterial devices is simply to miss half the picture." But then he adds that the category "technological societies" is best viewed as qualitative, not chronological. Such societies do not mark an epochal shift; nor do they point to a new stage in development. Rather, Barry argues, what is being defined here is a world marked by "[a] political preoccupation with the problems technology poses, with the potential benefits it promises, and with the models of social and political order it seems to make available. We live in a technological society, I argue, to the extent that specific technologies dominate our sense of the kinds of problems that government and politics must address, and the solutions that we must adopt. A technological society is one which takes technological change to be the model of political invention" (2001, 2).

Barry's observations are salutary. Epochal rhetoric can obscure how diverse peoples have always been enmeshed and entangled in complex social, ecological, and technological networks. Human history has always been entangled with the histories of diverse non-human agencies. Discussions of an emerging "third nature" to capture social change take a palimpsest-like approach to layers of natures. This suggests that "third nature" may well be too monolithic a concept for following the complex transformations and experiences of diverse technonatures occurring across the globe. Nevertheless, the social sciences are in the business of marking periods of change and continuity. As such, it might be said that the technonatural political-economy sensibility is marked by a growing desire not only to emphasize that humans have always been entwined in forms of material, technological, and informational networks but also that we are now witnessing a change of emphasis in modern technologies wherein—as Andrew Feenberg (1999)—observes, the transformations of the inner workings of natures and bodies have increasingly become more central to capitalist accumulation and technoscience. This has clearly increased the intensity and scale of the production of natures, and as Donna Haraway has observed, it forces us to address with renewed urgency the question *cui bono* ?—who benefits when these relationalities are transformed?

Ontology Matters

Transformations of political economy serve as important points of departure for many neo-Marxist and neo-Weberian technonatural inquiries. But those inquiries have often been supplemented—and of late sometimes supplanted in importance—by an emphasis on ontological matters. Many "technonatural conversations" inspired variously by Gilles Deleuze or Alfred North Whitehead, work in feminist science studies, and currents of actor network theory have argued that an attempt to grasp technonatural time/spaces requires as a first step a reworking and rethinking of the ontological domains of classical social theory. Approaches of this kind seem to share the view that neo-Malthusian environmentalisms and their contrarian detractors have for too long held to politics and theories implying that "Nature" either can be saved or, conversely, can be used as an infinite resource. Latour (2004), among others, has argued such approaches fail to recognize that political-ecological debates have never been about "Nature" in any straightforward sense as a realm beyond social and political life; rather, those debates have always been about the diverse ecological, technological, and social entanglements and connections we live *within*, the forms they take, and the interests those forms serve. A theme that often emerges in these ontological discussions is that we need to develop political ecologies that more explicitly recognize how our social-ecological worlds are actively made through diverse productions, circulations, and entanglements of people with other diverse non-humans, ecosystems, material and immaterial devices, and artifacts. Such currents suggest only through revising classical ontological orderings and practices will these developments be brought to light and will we understand the various subtle ways in which domination and control are presently being (often incompletely) made and remade in our "hybrid" or "more than human" worlds (Whatmore 2002).

To be sure, technonatural conversations that accent ontological matters are marked by different points of emphasis. But those conversations share a clear desire to move beyond past standoffs in the environmental social sciences, between realism and constructionism, beyond the purifications of modernist ontological separations of the natural, the social, and the technological. The allusive goal here is to develop a politics that can envisage our world as "real, material and discursive" (Latour 1993). Additionally, many of these currents seek to reclaim agency — human agency, to be sure, but also the agency of non-humans — so as to demonstrate that social-ecological worlds are not simply social constructs but are in a very materialist sense co-constructed. As such, this discussion resonates with the philosopher of science Karen Barad's (1996, 181–88; 2003) desire to formulate an "agential realism" based on performative intra-actions (not *interactions,* which presuppose distinct entities such as nature and culture acting on each other). The term "intra-actions" focuses on how reality is constituted in the "between"; in the inseparability of nature–cultural, world–word, physical–conceptual, material–discursive, so as to emphasize how humans, animals, materials, and things are not fixed prior to material discursive signification but in it.[4] It is in this "between" that many technonatural conversations want to be found.

What might we gain by such ontological reworkings? In many technonatural conversations one gets the distinct impression that reworking and retheorizing ontologies unearths a web of interconnections and intra-actions. In many technonatural ontological conversations one notices a strong desire to move beyond the flattening social constructionism of old, which presented reality as a smooth surface onto which social categories could be projected. In contrast, technonatural conversations are marked by a distinct preference for processual, dynamic, relational materialisms that can hold the "real" and "the symbolic" in tension and that acknowledge the "recalcitrance" of ecologies as well as the obduracy of objects.

On "Organic Nature," "Capitalist Nature," and "Technonature"

But do "technonatural" inquiries have to choose between political-economic and ontologically orientated approaches? Haraway—who has persistently and brilliantly combined the two approaches—would suggest not. Similarly, Escobar has combined insights from both approaches to propose a provocative framework for rethinking political ecology in technonatural time/spaces.

Escobar (1999) suggests that at the close of the twentieth century—an era of molecular technosciences, from recombinant DNA to gene mapping and nanotechnology—the view that nature is "untouched and independent" is increasingly giving way to "a new view of nature." He suggests that to investigate the diverse forms that the natural is taking in contemporary times, political ecology could do worse than analyze these developments within a framework that conceptualizes them in terms of three distinct but interrelated "natural regimes": "organic nature," "capitalist nature," and "technonature" (1999, 1). This framework would resemble Wark's evocation of "third nature" but with some important qualifications. Escobar stresses that within these regimes we need to be aware of (1) how nature(s) are differentially experienced according to one's social position—that is, they are differentially produced by divergent historical groups in different periods and spaces; (2) how these regimes "do not present a linear sequence or series of stages in the history of social nature"—rather, they "coexist" and overlap"; (3) how, moreover, these "nature regimes" produce and reproduce one another and are characterized by mutual "linkages and leakages"; and (4) how we need to constantly attend to the ways in which the resources for inventing (or we might better say, co-producing) these "nature regimes" are unequally distributed.

If we were to follow Escobar, we might see "the technonatural" as a "nature regime" interacting with organic and capitalist natures in complex and contingent ways. In this collection, though, we seek to expand this understanding of the technonatural. "Technonatures" in this book is not simply a material referent for emerging artificial natures; it is understood, just as much, as a cultural sensibility, a phenomenon of everyday life, an imaginative horizon, and an ideology.

For example, one can point to a range of imaginative phenomenologies related to the technonatural geographies and sociologies of everyday life (e.g., Michael 2000, 2006; Birke, Arluke, and Michael 2007; but see also Michael in this book; Clarke 2002; Thrift 2005). All of these studies seek to capture how natures and modern built environments have become increasingly interwoven with diverse people's everyday experiences. In contemporary culture, illustrative technonatural sensibilities seem to be hovering in dystopian cinematic visions such as *Gattica, I-Robot and eXistenZ;* in more hopeful gestures such as the ecological-technological art installations and spectacles of Olafur Eliasson or the more involved, open-ended boundary crossing experiments in visual culture, design, and geography by Kathryn Yusoff and Jennifer Gabrys; and in the rather more controversial new aesthetic interventions in biotechnology debates provoked by Eduardo Kac (see Kac 2007). Moreover, these more positive technonatural sensibilities seem to be seeping into activist circles. Consider the efforts by the Bioneers in California to draw together currents of Green activism with progressive engineering; and Brian Milani's ecomaterials project in Toronto, which seeks to bring together trade unionists with environmental activists and engineers to explore communalist and democratic expropriations of the diverse possibilities of the "new productive forces" from industrial ecology and post-Fordist ecological technology (see Milani in this volume). Or consider the ECO-TEC group, which proposes an ecologically sustainable architecture and posits that this must take the form of a "catalytic fusion" between "ecology and technology" (Marras 1999; see also Guy in this book). In all of these developments we can see the surfacing of a "technonatural" sensibility.

In this collection the metaphor "technonatures" names a moment during which the environmental debate seems to be folding into a vastly more complex social–ecological–technological field of political discussion. "Death of environmentalism" discourse focuses on a malaise, yet perhaps it is more accurate to say that the nature of the "environment" is being contested, expanded, *and* rendered plural. As such, environmental politics is necessarily becoming subject to extensive revisionism. All of the contributions to this book, in focusing on the city, culture, technologies, and body, and the future of politics and the environmental social sciences, point to these shifts.

Plan of This Collection

The collection begins with a range of chapters that highlight the various ways in which different traditions of environmental studies are defining, exploring, and critiquing material, cultural, experiential, and ideological manifestations of technonatures. We begin with a contribution from

Peter Oosterveer. Oosterveer's colleagues in the Environmental Policy Group at Wageningen have over the past decade developed some of the most influential and sophisticated renderings of the sociology of ecological modernization. By detailing the constructive role that state action and careful environmental policy making can play at the international and national levels in ameliorating environmental problems, this tradition of environmental sociology has operated as something of a Weberian foil to eco-Marxist sociology. In contrast to the "grow or die" or "treadmill of production" narratives that have been deployed by numerous Marxist scholars to present capitalist political economies as necessarily locked into an environmentally disastrous "grow at any cost" dynamic, ecological modernizers have pointed to a range of legislative, technological, and cultural developments — from the Montreal Protocol to the Dutch National Environmental Plan — which indicate that the capital/state/sustainability relationship could still play out in rather different ways. These approaches have provided some of the most interesting critiques of "radical ecological narratives" over the past decade. However, the manner in which ecological modernizing narratives have grounded their claims on nation-state–centred case studies — which often focus on the more progressive modes of environmental governance emerging out of Northern European countries — has opened this current to critique. A number of cross-national studies of ecological displacement and ecologically uneven exchange emerging out of world systems theory suggest that in certain places and spaces, it may well be that environmental improvement can be achieved only by exporting environment degradation to other spaces and places.

Oosterveer's chapter provides a useful summary of how the sociology of ecological modernization is repositioning itself to address the rise of both "global times" and "technonatural times." The environmental social sciences sometimes seem to inhabit a world in which conventional distinctions among the social, the natural, and the technological, and between space and time, are becoming blurred. In this regard, Oosterveer's chapter embraces a realist definition of environmental problems. The "technonatural condition" is framed in terms of new challenges for environmental governance that are emerging from globalization as more and more poorly regulated material flows — from waste and genetic material to natural resources and energy — cross a variety of transnational spaces. Oosterveer argues that what is required are new modes of analysis — specifically, the development of a "sociology of environmental flows" that can "map the movements of environmental problems." He maintains that this will "help identify more adequate modes of environmental governance." For theoretical inspiration, he renews the sociology of ecological modernization by drawing it into dialogue with the work of Manuel Castells, John Urry, and Arjun Appadurai on "networks" and "flows" and on material-flows analysis in "industrial ecology." To demonstrate some of the empirical complexities involved in developing this paradigm, he offers a case study of the global effects of shrimp farming.

Erik Swyngedouw is one of the most influential figures in contemporary urban geography. He promotes the value of exploring commonalities between historical-geographical materialism and hybridity narratives. In his chapter, he marshals together Marx, Lefebvre, Bruno Latour, and Donna Haraway to formulate a "cyborg urbanism." Rather like Oosterveer, Swyngedouw seeks to develop a dynamic and mobile conceptual framework for capturing the challenges posed by the emergence of technonatural time/spaces. But in contrast to Oosterveer, who turns to Castells and Urry for inspiration, Swyngedouw seeks to demonstrate how "cyborg" and "hybrid" discussions allied with Marx's interest in mapping circulations and metabolism open up the imaginative horizons of technonatural inquiries where it is not the hybrid that has ontological priority but the process of hybridization (Bakker and Bridge 2006, 17). He argues that such conceptual frameworks can bring into focus new objects and new relations between objects; but most crucially, such approaches can underscore the centrality of the city and the urban to the modern environmental debate. He suggests that a key challenge in future environmental discussions will be to understand how natures are becoming urbanized and how the urban and the city itself are best understood as complex social-ecological and social-technological processes. Swyngedouw contends that the emergence of technonatures cannot be viewed as universally negative or positive. Rather, an urban political ecology needs to attend to the complex power geometries that generate winners and losers in the processes of making and remaking the city and of the urbanization of nature. Put another way, urban political ecology needs to work less with general crisis narratives and attend more to the specifics of changing power relations.

Many eco-Marxists, and many sociologists of ecological modernization, approach technonatures through a macro- or at least meso-level perspective. In contrast, the contributions to this collection by Michael and by Hinchliffe and Whatmore highlight the virtues of situated and qualitative approaches to grasping "hybrid worlds."

Mike Michael, like de Certeau and Latour, draws inspiration from the sociology of everyday life. His very different approach to technonatures highlights how mundane objects, humans, animals, and miscellaneous artifacts are assembled, routinized, standardized, and ordered into diverse technonatures. In his chapter he extends these themes in order to explore the ironic spatialities of the cellphone in the British countryside.

In Britain, "the countryside" has long held a cherished place in many people's dreams of "getting away from it all." Yet what has always been a worked landscape is increasingly becoming an "electromagneticized countryside." Michael describes the many ways that cellphones have become mundane and the countryside thereby "domesticated." The chapter considers the various ironies at work here—generating the paradox that the everyday enactment of "nature domesticated" has generated a "technonature wild."

Sarah Whatmore's *Hybrid Geographies* (2004) and Steve Hinchliffe's *Geographies of Nature* (2007) have emerged as two of the most influential discussions of socio-natural hybridity to come from the discipline of geography in the past decade. In the closing contribution to this section, they provide us with a fine example of their unique post-human approach to technonatural studies. Drawing together insights from Gilles Deleuze, Bruno Latour, Isabelle Stengers, and actor network theory, they offer an account of urban political ecology that accents agency, heterogeneity, and specificity.

Hinchliffe and Whatmore open with several accounts of how some animals have adapted to the city—another aspect of becoming that might be seen as part of technonatural time/spaces. These stories illustrate the heterogeneity of the urban. They then switch focus to the "doing" of urban ecologies to suggest that in heterogenous cities, ecologists, social scientists, activists, and others do not unveil truths about the world so much as they intervene in the co-constructions of realities. Here, they are following Latour in suggesting that a social science that takes hybridity seriously requires new epistemological practices of doing research and politics—practices founded on open and contestable "matters of concern" rather than on closed, institutionally closed "matters of fact." For Hinchliffe and Whatmore, new forms of politics need to take seriously, engage, and invite into policy-making processes the various constituencies that constitute the "more than human" worlds in which we find ourselves enmeshed. Drawing from Stengers, they point to rather different styles of environmental politics—notably, a potential "politics of conviviality" that might unfold from this.

Technonatural Bodies, Cultures, and Everyday Life

In the second section of this book, the discussion of emerging technonatural worlds moves into the realm of culture, the body, and everyday life.

Julie Sze's chapter draws from literary analysis and discourses on environmental justice to examine one instance of how human and animal bodies interact with technological and environmental systems. Her chapter examines the case of diethylstilbestrol (DES), an artificial estrogen. As Sze documents, between four and six million women in the United States between 1948 and 1971 were prescribed DES to prevent miscarriages. She notes that it was also standard practice for farmers to use DES to fatten chickens, cows, and other livestock. However, DES proved to be toxic for both women and livestock. Informed by ethical concerns about environmental justice, and building on Haraway's articulation of the cyborg as a hybrid of organism and machine, Sze argues for a culturally grounded analysis of DES through the framework of "technologically polluted bodies."

Sze advances an understanding of technologically polluted bodies as hybrids of animal and human bodies—female bodies in particular. She then links these polluted bodies with non-machine-based forms of technological intervention, such as products developed and normalized by the pharmaceutical, petrochemical, and livestock industries. It is her view that an analysis of DES through the framework of technologically polluted bodies highlights how categories of race, gender, human/animal, and nature are unstable and are shaped and contested by ideas and cultures as well as by corporate industries that materially shape these categories—actively so—through their products and processes. "As a case study in polluted women and livestock (animal bodies)," she tells us, "DES illustrates changes in the human relationship to nature and what these changing relationships might mean for the possibility of justice and ethics in a hyperpolluted, highly technological world of corporate concentration." She offers a rich case study, one that could be complemented by many other instances of mass medication of human and animal populations during which unaccountable numbers of people and animals have served as guinea pigs.

Sze's discussion of the rise of specific technonatural cultures is pained. In contrast, Fletcher Linder focuses on the possibilities that might emerge if we recognized more clearly our diverse, embodied technonatural condition. Linder attempts to recover possibilities for sensuous engagement with, and a reclaimed politics for, technonatural space/times. To that end, he writes about a cultural practice that at first seems far removed from the environmental debate: bodybuilding. Drawing on fifteen years of ethnographic work on bodybuilders, he poses this question: How might the technonatural figure speak *productively* in environmental debates? By detailing the training practices and aesthetic logics of Southern California's elite bodybuilders, he locates the built body in its technonatural cultural and social contexts in order to specify two ways in which the built body poses useful lines of inquiry into environmental concerns. His first theme builds on a phenomenological exploration of training and nutrition to ask questions about the co-generation of body, world, and consciousness. He suggests that human entanglements with things enables us to envisage "the environment," "the body," and "subjectivity" as material products of ongoing and mutable metabolic processes. The result renders materiality as a product of interaction, craft, and sensuality; it also fractures simplistic and often debilitating distinctions between humans, knowledge, and a material world. The second line of inquiry highlights craft and making. Building on Elaine Scarry's recent work in aesthetics, he finds in bodybuilding a broad concern for an ethics of living. He attempts to show how an aesthetic and ethical approach to the built body might be a basis for thinking about technonatural environmental ethics.

Aidan Davison's chapter tries to steer a middle course through the theories of Hinchliffe, Whatmore, and Swyngedouw. His subject matter is technonatural conversations in Australia as these relate to suburbanization and the future of environmentalism. He examines how Australian environmental social movements are experiencing a loss of confidence, coherence, and direction, much like those in the United States and Europe. His work is of particular interest for the manner in which he reports feelings of "bad faith" and "guilty conscience" among Australian environmental activists. Indeed, he suggests, "participants feel acutely the disjuncture between their discourse and the world they inhabit." These activists are increasingly aware of how hard it has become to deploy the categories of pristine or pure nature in the complex, hybrid lifeworlds of their everyday lives.

Thinking Technonatural Present-Futures

In this book's final section we reflect further on how the challenges of technonatural time/spaces might soon begin to play out in environmental studies and among environmental social movements.

Tim Luke was one of the first scholars in the broad critical-theory tradition to argue the virtues of combining insights from the Frankfurt School, Foucault, and Haraway to grasp evolving "cyborg ecologies." In his chapter he posits an emerging technonatural future that is somewhat dark and foreboding. "Technonature" for Luke is best grasped as a new material environment emerging from what he refers to as today's "neo-capitalist" order. It is defined in part by the ever-present buildup of noxious by-products in and around urban industrial formations and in part by broader "unsustainable worldwide webs of economic exchange." As one would expect from a critical theorist, this understanding of technonature is developed in a macro fashion and grounded in historical analysis and political economy. Luke views technonature not just as an unfortunate and potentially catastrophic evolving material reality but also as an ideological construction. He warns of the ideological role that "technonatural talk" can play in reifying and reducing nature. How can we resist the spread of cyborg ecologies? While Luke recognizes the growing limitations of "green-grounded criticism," he cautions against embracing the solutions pointed to by the death-of-environmentalism narrative. Ultimately, he suggests that the challenges posed by emerging technonatures and by the malaise of conventional environmentalism can be countered only if we generate a much more extensive democratic politics—in his terms, a "public ecology."

Simon Guy develops thoughts on the complexities that hybrid worlds bring to conversations about the built environment by reflecting on the trials and tribulations that have emerged over the past two decades in formulating a form of green or sustainable architecture. He argues that technonatural narratives pose considerable problems for conventional forms of environmental architecture—problems that are rooted in essentialist notions of nature. But he goes on to suggest that a technonatural concern with hybridity and complexity could just as easily compel sustainable architecture to more carefully formulate more site-specific architectures.

In the final chapter, Brian Milani turns the tables on our discussions of changing environment–technology–society relations. While much techno-natural talk assumes there is a crisis in environmentalism, Milani suggests that a lot of the talk about "cyborg ecology," though it has some value, can actually obscure a rather bigger problem at the moment, which is the malaise of the political and intellectual left. Milani contends that though much of the academic left has been greatly agitated in recent times by the various "hybrid" developments being produced by biotechnology and other technosciences, this left has lost almost all interest in mapping the basic changes that have occurred in the character of contemporary productive forces. He argues that the lack of interest in these developments is somewhat ironic, since ecotechnological innovation and eco-industrialism embedded in a broader radically democratic community politics offers the best possibilities for rethinking a green political economy and a regenerative ecological politics. Clearly inspired by Murray Bookchin's call for an optimistic post-scarcity vision of an alternative socio-ecological vision (Bookchin 1971, 1980) and reflecting the growing convergence of environmental justice, ecotechnology and urban economic redevelopment that is formulating around talk of the "green collar economy" (Jones 2008), he speculates on the possibilities that exist for post-industrial and post-material forms to generate new modes of qualitative development.

Technonatural Conversations: Disputes, Divisions, and Possibilities

This collection demonstrates that a technonatural imaginary can open up new approaches and new ways of thinking about future political ecologies. Yet it also demonstrates that this conversation is at an early stage. There are many points of tension among the contributors to this volume—indeed, technonatural approaches are already amassing critics and skeptics (see Soper 2004; Benton and Craib 2004).

Hybridity has become the mantra of many technonatural conversations. Yet if hybridity is simply seen in terms of the assertion that everything is mixed or related to everything else, it clearly engenders the realist response that not only is this observation potentially trite but used in an undefined fashion the notion potentially prevents differentiation and hence collapses into one large mono-ism. Perhaps hybridity then does need to be used in more specific terms, notably as an appreciation that differentiation is inevitable and that while we persistently deploy (and perhaps need to deploy) the categories "natural," "social," "cultural," "technological," the evoking of hybridity seeks to draw attention to the sense that these categories are historically and geographically contingent and the differential boundaries between such fluid categories are contestable and power laden (Barry 2001), they leak and they change over, and across, differing time-spaces.

A further matter of dispute that underlies many of the chapters in this book concerns the question of who has, or indeed where is, agency in technonatural worlds. While all the contributors to this book seek to think about social and political life as folded and unfolded within an array of broader ecological and technological networks, there is a clear difference between the neo-Marxist cyborg approach of Swyngedouw (additionally, see Gandy 2002, 2004, 2005; Kaika 2004; Bakker and Bridge 2006), for example, which still locates agency in the world of the social, and the "more than human" approaches of Whatmore and Hinchliffe, who in post-humanist fashion are much more committed to developing hybrid accounts of the world that see multiple agencies distributed in networks with, and yet beyond, people (see Braun 2004). The claim that agencies are multiple and that they are distributed in "more-than-human worlds," as the chapters by Michael, Sze, and Hinchliffe and Whatmore demonstrate, can have real gains for political ecology. Yet if a clear danger of neo-Marxist approaches to technonatures is that "nature" and other non-humans can end up functioning as something of an inert backdrop to the story being told (Braun 2005; Hinchliffe 2007), a possible danger of "more-than-human" approaches is that the desire to discover multiple agencies, to re-animate the world beyond humans can, if it is not carefully done, end up both anthropomorphizing this world beyond humans and ensuring that a rather passive human subject fades perhaps too quickly into the general network. This can become particularly apparent in attempts to formulate a viable post-human politics, since it does seem to struggle with the observation that to talk of political ecology (and hence political agency) perhaps does require us to recognize at some level that we are still talking of people being active, of people potentially being the prime movers for political change, even if such changes will necessarily be in concert, made out of intra-actions with many other things as well.

Third, clear tensions exist in technonatural conversations between macro, meso, and micro orders of analysis. Andrew Barry suggests that reimagining the social (or socio-technical worlds) in terms of excessive use of the metaphors of network and flows can not only reinforce technological determinism (the technological network underpinning social life) but can also ignore the local specificities of network construction and give too much solidity to the constant shifting, working, and reworking of socio-technical networks (Barry 2001, 14–20). Such observations are mirrored in Hinchliffe and Whatmore's chapter, which suggests that emerging technonatural worlds need to be grasped in ways that pay more heed to local nuances and textures. That is how attempted orderings of everyday life by governments and other institutions or businesses are always "interfered" with in the spaces where they are enacted as they seemingly always jar with other (dis)ordering activities, and as such are to be analyzed as *attempted doings* that often fail in many of their goals or are incomplete and have many inadvertent effects (Hinchliffe and Bingham 2008). Other contributors, however — notably Luke, Swyngedouw, and Oosterveer — argue that while attention to specificity and particularity are needed, equally important is to recognize the relatively enduring qualities and power relations of some networks (i.e., capitalist political economies) over others and perhaps the enduring importance of maintaining a commitment to the notion that there are certain common logics (i.e., capitalist logics) to be found constantly working across networks — though again it is acknowledged that these may be worked in incomplete ways, resisted, or just do not work (see also Castree 2006; Braun 2006).

Technonatural Futures? Politics, Ethics, Alternatives in Technonatural Worlds

So if a central feature of technonatural sensibilities, in contrast to mainstream environmentalism, is that they recognize that one "nature" does not provide an ethical or political gold standard from which all else can be judged, that what we call "nature" is not necessarily degraded by being mixed with humans, and that social life has always been constituted through diverse foldings and unfoldings — of ecologies, cultures, technologies, and diverse nonhuman others — if all of that, then where does this leave our prescriptive worlds and agendas? Do technonatural political ecologies suggest the need for a broad reworking of environmental politics — indeed, for an opening up of a "post-environmental" terrain (Nordhaus and Shellenberger 2007)?

The differing diagnoses of the malaise of environmentalism discussed in this introduction are working at different registers with regard to possible future trajectories. And in this context, some emerging technonatural conversations do seem marked by endless complexification of the present or persistent deferrals of political commitment. Yet among these practices there are signs that the mood is turning toward reconstructive conversations about the possibilities for alternative technonatural politics and ethics, and even toward speculation on alternative technonatural futures. This is a messy area, but we would like to suggest that three currents seem to be emerging in wider debate and in this collection.

Perhaps the most familiar political manoeuvre that technonatural conversations presently gesture toward is to champion what might be called an institutional-democratic or procedural-democratic politics of technonatural change. The reasoning goes that if it is demonstrated that both the possible framing of environmental problems *and* the directionality of social–technological–ecological change are not given in the nature of nature, this opens the possibility for the technophobia/technophilia seesaw to be transcended by a democratic politics of the environment, science, and technology. Such commitments are now well known; they are clearly compatible with the general sensibilities of longstanding strands of Science and Technology Studies and well as with the literatures on environmental justice and social nature (such themes are advocated most explicitly in this book by Tim Luke and Erik Swyngedouw). Luke, for example, concludes his chapter with the claim that the necessary supplement to the rise of cyborg ecologies is to formulate a "public ecology" as counterpoint. Swyngedouw's chapter similarly concludes with a gesture toward democratization of the productions of urban natures. He suggests that such a manoeuvre could bring more sharply into focus the extent to which the environmental concerns of the affluent world are not necessarily the immediate environmental concerns of most of the world (wood-fuel pollution, lack of potable water, and so on).

A second perspective represented in this collection (perhaps most clearly by Michael, Linder, and Hinchliffe and Whatmore) — what might be called a "cosmopolitical" tendency in the technonatural zeitgeist — is less focused on institutional-democratic issues (though it is by no means antagonistic to such projects) and more an attempt to address, challenge, and alter the cultural imaginary/cultural pathologies of the dominant currents of mainstream Northern environmentalism. Linder's defence of a green cultural politics that is attentive to craft, sensuousness, and the making of nature is in many respects an attempt to counter what William Chaloupka (2003) has noted as a tendency among Greens to embrace the "dismal" and the self-image of "the saviour." As Chaloupka (2003, 147) notes, "saviours" of course "preach, denounce and prophesise," and what seems to be missing in that is the "irrepressible lightness and joy of being green" as well as any sense that politics might be thought — as it so often has been acted — with a buoyant sense of possibility, a delight in intellectual speculation about openings. Linder's inspiration for refashioning techno-natural aesthetics is clearly in debt to Nietzsche's and Foucault's calls for an ethics of the self. Hinchliffe and Whatmore, though, draw inspiration from Isabelle Stengers, whose cosmopolitics has sought to articulate an alternative sensibility of political ecology or an "ecology of practices."

Stengers's "cosmopolitics" moves us well beyond the Science Wars by arguing that what is central to the sciences is adventure, and adventure is learning through a "creative enterprise," learning from new types of encounters with non-humans that have developed in the sciences throughout history. It is only the naive, realist science warriors who ignore this central role of creativity in the sciences, yet that role is clearly central, since "science is made and transformed by learning new ways of allowing itself to be influenced by its 'objects' giving rise to new practices" (Paulson 2001, 112). More generally, Stengers argues that this openness to changing encounters can give rise to new political practices.

This kind of approach to political life, for Stengers, requires a new approach to taking risks, one that is very different from those which decry contemporary environmentalism and its supposed obsession with risk reduction. For Stengers, risk is more a practice of experimentation. It entails acting under conditions of uncertainty and constantly putting at risk "knowledge and assumptions of the knower." In her view, such an approach emphasizes the need to embrace the contingency of our world, a world of possibility, and the need to take risks to experience and change this world. Equally, though, she argues that it requires a new way of doing the human sciences: "The beautiful problem for human sciences, the problem for which we need a production of knowledge, is not what determines people or explains the way they think and feel, but the kind of processes which may transform weak isolated people into a collective able to invent its own position, its own strength" (Stengers in Zournazi 2002, 265).

Stenger's cosmopolitics poses many challenges for environmentalism. It could bring very different, more respectful relations with people, with publics, just as much as it should do with the diversity of non-human animals, so that people are not made into pretexts or objects to be scolded, told what to do, faced with fines or even guns for not doing what states decree through their environmental policies — policies that are focused on "aberrant individual consumption [but] have the power to make a difference and put what others think they know at risk" (Paulson 2001, 113), and in this process accept the risk and challenge of being changed in such encounters.

Of course, many environmentalists may find that this sort of experimentation, this risking of the knowledges they have felt so safe with (at least until recently), all too easily plays into the hands of their detractors. Even so, the cosmopolitical approach builds on how the historical view of sciences has been developing, and it does so in ways that point toward a broader possibility of politics between people and between people and non-humans. It is an ambiguous politics, but one that Hinchliffe and Whatmore set out in more empirically based ways than others before them, and it is suggestive of a new environmental politics.

A third and final current in progressive technonatural time/spaces is what might be called a critical-anticipatory politics of socio-eco-technical remaking, a politics that suggests that the "remaking of environmentalist adventures" may well require taking further risks still. Perhaps the central theme of this current is that if we seriously hope to generate diverse public discussions of viable alternative technonatural futures, this may involve moving out of the easy and risk-free comfort zone of concluding our technonatural discussions with the standard post-structuralist "defence of the political" or reiterations of the virtues of antagonism. Perhaps to really embark on Stengers's "adventure" — to be open to ways that include openness to risk — will entail engagement at much more concrete levels.

Some decades ago Murray Bookchin argued that the most solid basis for an attractive post-Malthusian social ecology likely involved developing an ecological politics that was attentive to the dangerous disruptions produced by contemporary capitalism *and* at the same time grasping the social, technological, and ecological potentialities that post-industrial and ecotechnological developments might already be able to offer post-scarcity futures. Bookchin maintained such a social ecology needed to attend to "desire" as well as "need" to envisage different socio-ecological futures (see Bookchin 1971, 1980; Heller 1999, White 2008). Bookchin's optimistic visions have perhaps fallen out of favour. Indeed, the idea that ecotechnologies might be recuperated into a broader set of "liberatory technologies" (Bookchin 1971) that might inform and compliment a political and cultural project seeking to achieve a profound socio-ecological remaking of society seems to have been replaced by a largely managerial world view. However, Brian Milani argues in this book that optimistic, disruptive engagements between various eco-inovations and social movement activists continue to percolate through a range of debates about industrial ecology and sustainable agriculture, the green industrial revolution, and aesthetics. In some political spaces, to be sure, some of this literature takes on the form of technological fixes. But not all developments in these fields can be reduced to this; furthermore, there is a serious "pleasure of invention" present in much of this literature and in wider activist practices, and serious thought is given to the possibility of constructing very different social-ecological-technological infrastructures (see Milani 2002; Gottlieb 2002; White 2002). Such currents scarcely register at present on the horizons of more theoretically based technonatural discussions. But it would seem to be the case that if the technonatural conversations that have begun in this book are serious about disrupting the stultifying hegemonic contrarian/neoliberal/Malthusian axis of the environmental debate, attention needs to be paid to these conversations as well.

Perhaps we need to see more adventure and risk taking *among* the democratic, cosmopolitical, and critical anticipatory technonatural currents we discuss here. Because, as Isabella Stengers suggests, the adventure of thinking is ultimately an adventure of hope. And rather than crisis, perhaps more than ever what the environment–society–technology debate now needs is a large injection of social hope. As such, we also hope readers will find this emerging in the chapters that follow.

Notes

1 For a review of the Lomborg debate and a broader review of the rise of "anti-environmental politics," see White, Wilbert, and Rudy (2007).

2 We use the term "Northern" here recognizing that this term and "Southern" are problematic in that they seem to unduly specify a sense of unity within them and difference between them that masks wider differences and commonalities.

3 Holism has also a somewhat problematic and contradictory history with in the science of ecology and wider environmentalism. On an aspect of this see Anker (2001).

4 For Barad, the term "discursive signification" refers to specific materialist discursive practices. These discursive practices are: "(re)configuring of the

world through which local determinations of boundaries, properties, and meanings are differentially enacted. That is, discursive practices are ongoing agential intra-actions of the world through which local determinacy is enacted within the phenomena produced" (Barad 2003: 820–21).

Works Cited

Agarwal, A., and S. Narain. 1991, "Global Warming in an Unequal World: A Case of Environmental Colonialism." *Earth Island Journal* (Spring): 39–40.

Agyeman, J. 2005. *Sustainable Communities and the Challenge of Environmental Justice.* New York: New York University Press.

Anker, P. 2001. *Imperial Ecology: Environmental Order in the British Empire, 1895–1945.* Cambridge, MA: Harvard.

Bakker, K., and G. Bridge. 2006. "Material Worlds? Resource Geographies and the Matter of Nature." *Progress in Human Geography* 30, no. 1: 5–27.

Barad, K. 2003. "Posthumanist Performativity: Toward an Understanding of How Matter Comes to Matter." *Signs* 28, no. 3: 801–31.

———. 1996. "Meeting the Universe Halfway." In *Feminism, Science, and the Philosophy of Science,* ed. L. Nelson and J. Nelson. Dordrecht: Kluwer. 161–94.

Barry, A. 2001. *Political Machines: Governing a Technological Society.* London: Athlone.

Bauman, Z. 2000. *Liquid Modernity.* Cambridge: Polity Press.

Beck, U. 1992. *Risk Society: Towards a New Modernity.* London: Sage.

Benton, T., and I. Craib. 2001. *Philosophy of Social Science: Philosophical Issues in Social Thought.* London: Palgrave.

Biersack, A., and J.B. Greenberg. 2006. *Reimagining Political Ecology.* Durham, NC: Duke University Press.

Birke L., A. Arluke, and M. Michael. 2007. *The Sacrifice: How Scientific Experiments Transform Animals and People.* West Lafayette, IN: Purdue University Press.

Bookchin, M. 1971. *Post-Scarcity Anarchism.* San Francisco, CA: Rampart Books.

———. 1980 *Towards an Ecological Society.* Montreal: Black Rose Books.

Braun, B. 2006. "Towards a New Earth and a New Humanity: Nature, Ontology, Politics." In *David Harvey: A Critical Reader,* ed. N. Castree and D. Gregory. Oxford: Blackwell.

———, and N. Castree, eds. 1998. *Remaking Reality: Nature at the Millennium.* London: Routledge.

Bullard, R. 1990. *Dumping in Dixie: Race, Class and Environmental Quality.* Boulder, CO: Westview.

Castells, M. 1996. *The Rise of the Network Society.* Vol. I of *The Information Age: Economy, Society, and Culture.* Oxford: Blackwell.

Castree, N. 1995. "The Nature of Produced Nature." *Antipode* 27, no. 1: 12–48.

———. 2006. "A Congress of the World." *Science as Culture* 15, no. 2: 159–70.

Chaloupka, W. 2003. "The Irrepressible Lightness and Joy of Being Green: Empire and Environmentalism." *Strategies* 16, no. 2: 147–62.

Clark, N. 2002. "The Demon Seed: Bioinvasion as the Unsettling of Environmental Cosmopolitianism." *Theory, Culture, and Society* 19: 101–25.

Cronon, W., ed. 1995. *Uncommon Ground: Rethinking the Human Place in Nature.* New York: W.W. Norton. 69–90.

Crutzen, Paul J., and Eugene F. Stoermer. 2000. "The 'Anthropocene.'" *IGBP Newsletter* 41, May 2000.

Davis, M. 1998. *Ecology of Fear: Los Angeles and the Imagination of Disaster.* New York: Metropolitan.

———. 2002. *Dead Cities.* New York: New Press.

Deleuze, G., and F. Guattari. 1988. *A Thousand Plateaus.* London: Athlone Press.

Denevan, W.M. 1992. "The Pristine Myth: The Landscape of the Americas in 1492." *Annals of the Association of American Geographers* 82, no. 3.

Escobar, A. 1999. "After Nature: Steps to an Anti-Essentialist Political Ecology." *Current Anthropology* 40, no. 1: 1–30.

Feenberg, A. 1999. *Questioning Technology.* London: Routledge.

Frank, A. 1998. *ReOrient: Global Economy in the Asian Age.* Berkeley, CA: University of California Press.

Forsyth, T. 2002. *Critical Political Ecology: The Politics of Environmental Science.* London: Routledge.

Franklin, S. 2000. "Life Itself: Global Nature and the Genetic Imaginary." In *Global Nature, Global Culture*, ed. S. Franklin, C. Lury, and J. Stacey. London: Sage.

Gandy, M. 2002. *Concrete and Clay — Reworking Nature in New York City.* Cambridge, MA: MIT Press.

———. 2004. "Rethinking Urban Metabolism: Water, Space, and the Modern City." *City: Analysis of Urban Trends, Culture, Theory, Policy, Action* 8, no. 3: 371–87.

———. 2005. "Cyborg Urbanization: Complexity and Monstrosity in the Contemporary City." *International Journal of Urban and Regional Research* 29, no. 1: 26–49.

Gottlieb, R. 2002. *Environmentalism Unbounded: Exploring New Pathways for Change.* Cambridge, MA: MIT Press.

Guattari, F. 2000. *The Three Ecologies.* Trans. Ian Pindar and Paul Sutton. London: Athlone.

Guha, R., and J. Martinez-Allier. 1997. *Varieties of Environmentalism: Essays North and South.* London: Earthscan.

Haraway, D. 1991. *Simians, Cyborgs, and Women.* London: Free Association.

———. 1998. *Modest_Witness@Second_Millennium.FemaleMan©_Meets_ Onco Mouse ™.* London: Routledge.

————. 2007 *When Species Meet*. Minneapolis: University of Minnesota Press.

Hardt, M., and A. Negri. 2001. *Empire*. Cambridge, MA: Harvard University Press.

Harvey, D. 1996. *Justice, Nature, and the Politics of Difference*. Oxford: Blackwell.

Heller, C. 1999. *Ecology of Everyday Life: Rethinking the Desire for Nature*. Montreal: Black Rose Books.

Hinchliffe, S. 2007. *Geographies of Nature: Societies, Environments, Ecologies*. London: Sage.

————, and N. Bingham. 2008. "Securing Life: The Emerging Practices of Biosecurity." *Environment and Planning A* 40: 1534–51.

Hobson, K. 2007. "Political Animals? On Animals as Subjects in an Enlarged Political Geography." *Political Geography* 26: 250–67.

Kac, E., ed. 2007. *Signs of Life: Bio Art and Beyond*. Cambridge, MA: MIT Press.

Kaika, M. 2004. *City of Flows: Modernity, Nature, and the City*. London: Routledge.

Kull, A. 2002. "Speaking Cyborg—Technonatures and Technocultures." *Zygon* 37, no. 2.

Lash, S. 2001. "Technological Forms of Life." *Theory, Culture, and Society* 18: 105–20.

Latour, B. 1993. *We Have Never Been Modern*. Brighton: Harvester-Wheatsheaf.

————. 2004. *Politics of Nature*. Cambridge, MA: Harvard University Press.

Lomborg, B. 2001. *The Skeptical Environmentalist*. Cambridge: Cambridge University Press.

Luke, T. 1996. "Liberal Society and Cyborg Subjectivity: The Politics of Environments, Bodies, and Nature." *Alternatives* 21: 1–30.

————. 1997. *Eco-Critique: Contesting the Politics of Nature, Economy, and Culture*. Minneapolis: University of Minnesota Press.

————. 1999. *Capitalism, Democracy, and Ecology*. Urbana: University of Illinois Press.

MacNaghten, P. 2003. "Embodying the Environment in Everyday Life Practices." *Sociological Review* 51: 62–84.

Marras, A., ed. 1999. *ECO-TEC: Architecture of the Inbetween*. New York: Princeton Architectural Press.

Marx, L. 1964. *The Machine in the Garden: Technology and the Pastoral Ideal in America*. New York: Oxford University Press.

Michael, M. 2000. *Reconnecting Culture, Technology, and Nature*. London: Routledge.

————. 2006. *Technoscience and Everyday Life: The Complex Simplicities of the Mundane*. Maidenhead, Berks.: Open University Press/McGraw-Hill.

Milani, B. 2002. *Designing the Green Economy: The Post-Industrial Alternative to Corporate Globalization*. Lanham: Rowman and Littlefield.

Mol, A.P.J. 2003. *Globalization and Environmental Reform*. Cambridge, MA: MIT Press.

Nordhaus, T., and M. Shellenberger. 2007. "Break through." In *The Death of Environmentalism to the Politics of Possibility*. New York: Houghton Mifflin.

Paulson, W. 2001. "For a Cosmopolitical Philology: Lessons from Science Studies." *Substance* 96, 30, no. 3: 101–19.

Peet, R., and M. Watts. 1996. *Liberation Ecologies: Environment, Development, Social Movements*. New York: Routledge.

Philips, M., and T. Mighall. 2000. *Society and Exploitation through Nature*. London: Prentice Hall.

Philo, C., and C. Wilbert, eds. 2000. *Animal Spaces, Beastly Places: New Geographies of Human Animal Relations*. London: Routledge.

Rocheleau, D., B. Thomas-Slayter, and E. Wangari. 1996. *Feminist Political Ecology: Global Issues and Local Experience*. New York: Routledge.

Shellenberger, M., and T. Nordhaus. 2005. "The Death of Environmentalism: Global Warming Politics in a Post-Environmental World." *Grist* [online] January 15, http://www.grist.org/news/maindish/2005/01/13/doe-reprint.

Smith, N. 1984. *Uneven Development: Nature, Capital, and the Production of Space*. Oxford: Oxford University Press.

Soper, K. 1995. *What Is Nature: Culture, Politics and the Non-Human*. Oxford: Blackwell.

Stengers, I. "A 'Cosmopolities': Risk, Hope, Change." In Zournazi, M. (ed.) *Hope: New Philosophies for Change*. London: Pluto Press.

Swyngedouw, E. 1996. "The City as a Hybrid—On Nature, Society and Cyborg Urbanisation." In *Capitalism, Nature, Socialism*, Vol. 7(1), no. 25 (March): 65–80.

Thrift, N. 2005. *Knowing Capitalism*. London: Sage.

Urry, J. 2007. *Mobilities*. Cambridge: Polity Press.

———. 2000. *Society Beyond Societies: Mobilities for the 21st Century*. London: Routledge.

Wark, M. 1994. "Third Nature." *Cultural Studies* 8, no. 1.

White, D.F. 2002. "A Green Industrial Revolution? Sustainable Technological Innovation in a Global Age." *Environmental Politics* 11, no. 2.

———. 2008. *Bookchin: A Critical Appraisal*. London: Pluto Press.

White, D.F., A. Rudy, and C. Wilbert. 2007. "Anti-Environmentalism: Promethianism, Contrarians, and Beyond." In *Handbook of Environment and Society*, ed. J. Pretty et al. London: Sage.

Williams, Raymond. 1973. *The Country and the City*. Oxford: Oxford University Press.

Part One

Conceptualizing Technonatural Time/Spaces

Chapter One

Governing Environmental Flows: Ecological Modernization in Technonatural Time/Spaces
Peter Oosterveer[1]

While until recently the problem of environmental governance could be formulated in terms of a discrepancy between a globally organized ecosystem and nation-state–based regulatory arrangements, this no longer seems adequate (Young 1994, 2000). As other contributors to this book will also observe, the environmental social sciences find themselves in changing times as conventional distinctions between society and environment, the national, the local, and the global, between nature and technology, and between time and space seem to become blurred. In this chapter I want to draw these broad "technonatural themes" into productive engagement with two bodies of literature — the sociology of ecological modernization and the sociology of networks and flows — to reflect on the dilemmas of environmental governance in changing times. My starting premise is that many contemporary environmental problems related to waste, (green) products, genetic material, natural resources, and energy are most productively interpreted as poorly regulated material *flows* blurring time and place as they travel through a diversifying array of transnational spaces. I want to suggest that a theory of environmental flows crisscrossing national borders through transnational infrastructures and networks might help identify more adequate modes of environmental governance.

This chapter presents some building blocks for such a "sociology of environmental flows." The next section reviews how the concept of flows is applied in various environmental sciences; I also consider how the environment and changes in technology and society have been conceptualized within ecological modernization theory. The section after that outlines a sociology of global environmental flows. I indicate here the ways in which this approach can address current shifts in the organization of time and space and the related interactions among society, technology, and nature (or the environment). The (additional) value of a sociology-of-flows perspective is assessed and illustrated in the fourth section, which reviews the global governance of shrimp provisioning. The concluding section looks at transnational biodiversity management in response to "invasive species."

Theoretical Background: Different Sciences on "Flows" and "Governance"

The analysis of environmental flows has begun to take centre stage in the environmental sciences. In models of global climate change from the Intergovernmental Panel on Climate Change, for example, the concept of "flows" is a critical component of general circulation. Such "technical" studies arise from a natural-sciences perspective that applies concepts such as environmental systems, material-flows analysis, and industrial ecology (Ayres et al. 1996; Graedel and Allenby 1995; Spangenberg et al. 1998). In these studies the flows are taken as the central units of analysis to establish the linkages between material causes and effects. However, the social dimensions tend to remain undertheorized. The contribution of sociological analysis to such studies consists primarily in examining the institutions that "govern" these material flows. For example, it can be learned from the literature on global environmental governance in political science and international relations (see, for example, Keohane and Nye 1977; Krasner 1983; Young 1994) that nation-states are becoming less dominant while other organizational arrangements emerge and that new social actors are increasingly involved in managing environmental problems. What is largely missing from these studies, though, is a meticulous understanding of the broader social changes that are underpinning the shift from government to governance and the linkages between environmental governance and environmental flows. While several scholars of governance have identified the first challenge (Lipschutz and Conca 1993; Wapner 1996; Litfin 1998; Shaw 2000; Stevis and Assetto 2001; Stevis and Bruyninckx 2006), neither that challenge nor the second one has yet been adequately addressed.

In contrast to the natural sciences, the environmental social sciences have a long history of combining analyses of natural/material and social phenomena. Early strands of the Chicago School's human-ecology approach used the concepts and tools of plant ecology to make sense of human settlement patterns in 1930s Chicago. After these attempts, however, the physical/material reality disappeared from sight in sociology. Functionalism became the dominant school, and it demanded that social facts be explained solely by (other) social facts. It was only with the critique of this human exemptionalism paradigm (HEP), and with efforts to construct a new ecological paradigm (NEP) that evolved in American environmental sociology in the 1970s and 1980s, that efforts were renewed to bring the physical and material dimensions back in (Catton and Dunlap 1978). The NEP echoed the human-ecology tradition in its calls that social systems not be analyzed in isolation from physical flows. This paradigm discussed dependencies between human societies and ecosystems utilizing the Darwinian concept of the "web of life," which itself was framed in the neo-Malthusian language of physical limits to economic growth.

However, Allan Schnaiberg in his classic study (Schnaiberg 1980) on environmental sociology arrived at a conclusion that differed from that of Catton and Dunlap. Schnaiberg in his treadmill-of-production perspective argued against the partial or total fusion of the natural and the social. He also argued against fusing the disciplines designated for studying these different realms, since the social — so he contended — is different from the natural in some crucial respects. Societies are to be regarded as "dependent from" the sets of ecosystems they rely on for their proper functioning, but they do not function in the same (mechanistic) ways as ecosystems. Schnaiberg argued that since the social is different from the natural, the sciences of ecology and sociology should be kept separate as well, rather than fused in some human-ecology approach. The rise of a world-system tradition in environmental social sciences (see Bunker 1996; Goldfrank, Goodman, and Szasz 1999; Roberts and Grimes 2003) has placed social practices, institutions, and actors at the centre of analysis. However, this approach is also limited in that it analyzes environmental (especially energy) issues without investigating the environmental flows themselves in any detail. The result of this clear separation between society and ecosystems and of the focus on social processes is that the environment is conceived as the (passive) recipient of societal dynamics, such as industrialization and capitalism. As a consequence, only societal change is regarded as the appropriate solution to environmental problems.

In contrast to the treadmill-of-production and world system theory perspectives, ecological modernization theory (EMT) sets out to bring the social and material views in environmental policy-making closer together. EMT was developed in the late 1980s and the 1990s to account for the changes beginning to develop in the institutions and social practices involved in environmental deterioration and reform, especially in Germany, the Netherlands, and Denmark (Spaargaren et al. 2006b). The theory analyzes how environmental interests and considerations make a difference when processes of production and consumption are transformed in tandem with the institutions involved in their organization and management (Mol and Spaargaren 2002). Environmental flows between society and nature are conceived in terms of additions and withdrawals and analyzed in terms of their interactions with social dynamics, institutional arrangements, and policy structures. From its inception, EMT displayed a particular interest in the material dimensions of environmental change. EMT scholars share three broad perspectives:

- They attempt to move beyond apocalyptic orientations and to approach environmental problems as calls for social, technical, and economic reform.
- Instead of viewing environmental problems as immediate consequences of industrialization, they explore transformations of core social institutions of modernity — including science and technology, production and consumption, politics and governance, and the "market" — on multiple scales (local, national, global) in order to elucidate their impacts on the environment.
- Their position in the academic field is distinct from counterproductivity/deindustrialization, postmodernist/strong social constructivist, and many neo-Marxist analyses (Mol and Sonnenfeld 2000, 5). They focus less on physical environmental improvements per se than on social and institutional transformations that include environmental considerations.

Recently, efforts have been made to reformulate EMT in the context of both globalizing modernity (Mol 2001) and changing state–society relations (Mol and Buttel 2002). These innovations in EMT are intended to better accommodate the globalization process by establishing a more abstract, institutional perspective on environmental policy-making, governance, and the changing role of nation-states. These changes are also useful because they accord reasonably well with recent work in environmental-flows analysis. EMT posits that we are seeing the increasingly autonomous institutionalization of the environment. This autonomy of "the environmental" is reinforced through public as well as private decision making facilitated by the principles of decentralization and related arrangements. For instance, more and more private companies are developing their own internal environmental-management systems independently of official guidelines; meanwhile, many NGOs are promoting more environmentally friendly products through private labelling schemes. From an environmental-flows perspective, some observers take all of this one step further by suggesting that nation-state–driven policy-making has limited or bounded significance and by de-emphasizing the importance of global-regime governance (see Beck and Sznaider 2006; Buttel, Spaargaren, and Mol 2006). EMT and global-environmental-flows perspectives are sharing a disjunctural–transformational perspective when they declare that the late twentieth to early twenty-first centuries represent a watershed in the development of industrial capitalism. Finally, both perspectives emphasize that consumption *and* production are relevant when environmental problems are being investigated, and they share a commitment to "non-apocalyptic" views of environmental quality.

Developing a Sociology of Global Environmental Flows

Building on these considerations, I want to argue that a sociology of global environmental flows incorporating insights from the sociology of ecological modernization may offer an interesting starting point for analyzing environmental problems in global modernity (see Spaargaren, Mol, and Buttel 2006a, 2006b). Such a project can profit from engaging with the work of Manuel Castells (1996, 1997, 1998, 2004), Arjun Appadurai (1996), and John Urry (2000, 2003). All of these contemporary sociologists place networks and flows, rather than states and societies, at the centre of their analyses. What would seem to follow from their work is that a sociology of global environmental flows should address the relation between material flows and the social institutions and actors that govern them. The analysis of global environmental flows, then, should study not only the material substances and technical infrastructures but also the "scapes" (see below), nodes, societal networks, and discourses that go along with the flows in question.

The recent researches of Castells, Appadurai, and Urry offer many complementary insights for thinking about flows. Castells conceptualizes the transition to global modernity as the coming about of a global network society, one in which physical space is becoming less important and global flows are becoming more important. He views this global network society as a new way of structuring time and space by reintegrating the functional unity of different elements at distant locations made possible by modern transportation, information, and communication technologies. He argues that the configuration of these various networks has no isomorphic character; these networks differ in size and in the density of their networked connections as well as in their links with other networks. So some networks may be fundamentally fractal—without clear boundaries, structures, or regularities—while still overlapping with other networks. For example, Urry (2003) distinguishes between global integrated networks (for which multinational companies such as McDonald's constitute the stereotypical example) and global fluids, which are much more liquid in character (such as global media). Global networks are stretched across multiple and distant spaces and times that are no longer closely bound (Adam 2000). Biological time, the primary rhythmic organizing principle during most of human existence, changed to clock time in the industrial age, but today we live in the age of "timeless time" (Castells 1996). In this "timeless time" the normal sequencing of events is beginning to disappear: time is becoming self-maintaining, random, and incursive, thus breaking down familiar rhythmicity. Space, too, is becoming largely irrelevant, with exchanges and social interactions increasingly taking place without face-to-face contact. Interactions among people, machines, texts, objects, and technologies are occurring across multiple and distant times and spaces mediated by various material worlds (Urry 2003). Positive as well as negative feedback mechanisms may be distantiated in terms of the time and space from which they originate. To explain the dynamics of these configurations, we must rely on images of flows and uncertainty, dynamics and irreversible change, rather than on older images of order, stability, and "system-ness." When flows of materials, capital, information, images, and so on that connect physically disjointed positions gain some permanence, Castells speaks of the "space of flows."[2] A space of flows may be largely virtual, or it may have more material specificity when it concerns movements of matter and energy. Global flows travel at high speed across national borders, along various "scapes" (Urry 2003). This notion of scapes relates to the varying situatedness of the actors engaged in constructing the global flows; at the same time it points to their fluid, irregular shapes. Because each scape is subject to its own constraints and incentives, the emerging global system is not necessarily coherent; rather, it is often a set of disjunctive dynamics (Appadurai 1996). Environmental flows are a concrete example of these disjunctures and complexities, which are especially visible in the problems facing global environmental politics.

When we incorporate some of these insights of the sociology of flows into ecological modernization theory, it is clear that the linear metaphor of scales of government, from micro to macro level, or from life world to system world, should be replaced by the metaphor of governance connections. There is not a one-dimensional process of transference of power from local and national levels to international networks; rather, there is a global dynamic with contradictory tendencies regarding which the outcome cannot be known beforehand. Thus it becomes clear that a revised environmental sociology should not focus on the characteristics of fixed entities, on localities, but on the relations among them and the resulting global complexity. The complexity of transnational flows prevents individual human action from having a determining influence on the direction and structuration of their movements; yet at the same time this offers opportunities for interventions along the flow as opposed to solely from the unique central position of governments. Some pockets of ordering, or global properties, may emerge, but these are not generated by linear effects because of the presence of global complexity and because causes are always overflowing from domain to domain and especially across the supposedly distinct and purified "physical" and "social" domains (Urry 2003). The global is not to be situated as opposing the local, because global flows help produce the local and vice versa. The global and the local are inextricably and irreversibly bound together through a dynamic relationship, and connections between the two are to be viewed as more or less mobile, more or less intense, more or less social, and more or less "at a distance" (Beck 1997).

A sociology of global environmental flows should therefore analyze the material and social characteristics of such flows and the ways in which they travel along different scapes. The local and global environmental impacts of these flows should be assessed, and the key social actors in their governance should be identified. Environmental flows include transnational and local material dimensions as well as informational, conceptual, and monetary aspects. They exhibit particular tensions between the global space of flows and the local space of places, between the virtual and material dynamics, and between the different components of the networks and scapes. An exploration of how transnational environmental flows could be governed should thus be based on a consistent analysis of their internal dynamics and external relations. National governments remain important actors in the national and international governance arrangements that are being developed, but they are no longer necessarily the key actors in this. They are losing their traditional meaning as "a sovereign state, whose hierarchically imposed commands are binding on all parties subject to its jurisdiction, while at the international level decisions are taken by sovereign states acting unilaterally or through formal or informal modes of inter-sovereign co-operation" (Karkkainen 2004, 76). In a countervailing fashion, nation-states are transferring selected powers to supranational agencies and institutions such as the European Union (EU) and the World Trade Organization (WTO). However, despite these efforts and the presence of exceptional levels of global interdependence, it is still so far improbable that a unified global state will replace the various national ones and deal effectively with existing global problems. The sociology-of-flows perspective makes it clear that government authorities have to share their role in governance with many other social actors linked into global networks and scapes (Held et al. 1999). The involvement of these other, non-state social actors may result in the emergence of innovative and more flexible forms of global environmental governance, signifying the "fluidization" of regulatory practices (Lipschutz and Fogel 2002). In the age of globalization, the "gardening state" (Bauman 1987) of simple modernity—which is chiefly concerned with patterns, regularity, order, and control—is increasingly unable to fully control domestic social processes and merely regulates (the conditions for) mobility. Every state has to transform itself into a network-state—that is, become part of a complex web of power-sharing and negotiated decision-making processes involving transnational and national political institutions as well as market actors and NGOs.

The perspective of a sociology of flows introduces an approach to environmental governance that can address several of the issues mentioned.

In its analysis of environmental problems, local and global dynamics as well as social and material dynamics are linked continuously. Through this focus on global environmental flows it becomes possible to analyze the involvement of various (governmental as well as non-governmental) social actors in this flow. These social actors do not necessarily have to be located in the same places, and they may not even be in direct contact with one another. Nevertheless, these social actors interact with the material flows, and through a combined analysis of both dynamics it becomes possible to identify options and difficulties in establishing more adequate environmental-governance arrangements.

The cases of the global supply of shrimps produced from aquaculture and the global flow of invasive species, presented in the next two sections respectively, will illustrate this perspective further.

Shrimp: Governing Flows of Food in the Context of Globalizing Production–Consumption Chains

Thai shrimp form one of the many flows of food travelling around the world along different scapes, with increasing speed and volume, over ever larger distances and in more and more different directions (Fine 1998; Goodman and Watts 1997; Goodman 2001). Global flows of food emerge as food processors and retailers in Western countries source more and more of their food globally. Agri-food flows link the places of production with the places of consumption and involve various social actors apart from the producers and the consumers, such as traders, processors, and retailers, and also governments and NGOs (Oosterveer 2005, 2007). The lengthening of these flows is intimately linked to the progressive industrialization of food, whereby natural processes are replaced with/substituted by industrial processes, so that food products are transported over longer distances, thereby influencing the lives of both producers and consumers (Beardsworth and Keil 1997). These flows cover a whole variety of social, technical, economic, and natural components (Murdoch 2000), which are characterized by time/space distantiation and reproduced within contextualized social practices (Giddens 1979; Van der Meulen 2000; Dicken et al. 2001). Contrary to, for example, virtual flows such as global finances, global food supply can never become completely disembedded from place and time, or "footloose." The governance of global food flows therefore needs constantly to accommodate the dynamics of the global space of flows and the local space of places.

The space of flows of shrimp not only includes the crustacean itself, but also the related information and finances (see Figure 1.1). So it is incorrect to regard shrimp production and consumption as consisting of a one-directional chain of supply covering the successive steps in production, processing, and trade. In the space of flows, multidirectional exchanges occur among different actors, immaterial aspects, and material objects.

Shrimp currently account for 20 percent of the total value of internationally traded fish, making it the most important fish commodity in value terms. Nearly 80 percent of the world's production enters the global market (Goss, Burch, and Rickson 2000). In 2003 world shrimp production reached a total of 4.2 million tonnes, of which some 1.6 million (38 percent of the total) was from aquaculture, or shrimp farming, compared to about 5 percent in the early 1980s (Josupeit 2004). Shrimp farming represents a total value of US$6.9 billion at the farm gate and $50 to 60 billion at the point of retail. These shrimp farms are mostly located along rivers, estuaries, and coastal areas in tropical regions. The leading producers are China, India, and Indonesia, but Thailand is the largest exporter by far, exporting some 90 percent of its total output and supplying about 18 percent of the world market for frozen shrimp.

It is not surprising to see widespread shrimp farming in Thailand: the country has the optimal climate for this activity, a coastline of about 2600 kilometres, and a well-developed infrastructure. Extensive shrimp farming with low yields began in 1935 along the eastern coast of the Gulf of Thailand in rice fields, where shrimp were harvested both for private consumption and for sale on the local market. Relatively good prices and

FIGURE 1.1

Global Flows of Shrimp Connecting Production with Consumption

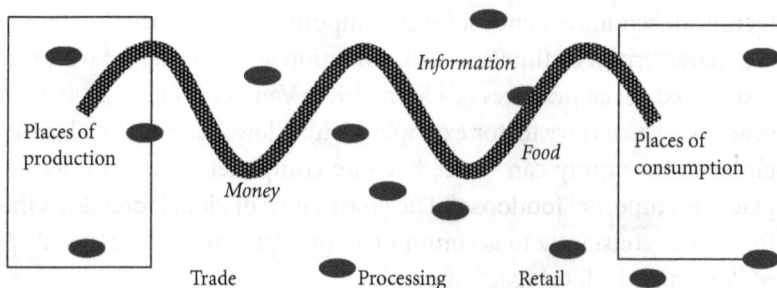

severely reduced salt prices in 1947 created the conditions for an expansion of extensive shrimp farming. When in 1987 Taiwan's intensive shrimp industry collapsed, Thailand quickly filled the gap. Intensive shrimp farming is based on high stocking densities, whereby all shrimp are supplied by hatcheries, processed feed is used, water is often flushed, and farming is mechanized (aerators, water pumps, lighting). However, intensive shrimp farming regularly leads to disease outbreaks in the ponds caused by viruses, bacteria, fungi, and other pathogens (Gräslund and Bengtsson 2001). So repeatedly, Thai shrimp farmers have had to look for new, disease-free locations. Initially, shrimp farming was located in the inner gulf (Bangkok area), but via the Western Gulf, the Eastern Gulf, and the Andaman Sea, shrimp farming is now moving inland (Huitric et al. 2002). Simultaneously, farm-management practices have been adapted, by replacing open-culture systems with semi-closed recycling systems in which pond water is treated after each crop and then reused (Flaherty and Vandergeest 1998; Duraiappah et al. 2000).[3] Currently around 85 percent of Thailand's shrimp are produced by intensive farming and only 15 percent still by extensive or semi-intensive farming methods. Commercial shrimp farming has become highly complex, involving thousands of small, family-based enterprises, traders, and processors as well as retailers from all over the world and supermarkets, hypermarkets, and convenience stores.[4] Thailand's dependence on global markets makes the country rather vulnerable to changes (in the national economic situation but also in the domestic regulatory practices) in Japan, the United States, and the EU. Despite this vulnerability, because of the largely unregulated character of the shrimp supply chain, the opportunities for high and quick returns on investments continue to make the shrimp aquaculture a highly attractive industry.

Thailand's booming shrimp industry is also generating serious problems. These include environmental damage (Pillay 1992; Folke et al. 1998; Naylor et al. 2000; Goldberg, Elliott, and Naylor 2001), food-safety risks (Benbrook 2002; FAO 2002), and negative socio-economic effects for the coastal populations (Maybin and Bundell 1996). Shrimp-pond construction often involves clearing mangrove forests that protect coastlines and that serve as nurseries for other fish species. Furthermore, the construction of these ponds may contribute to serious social conflicts over land rights and access to natural resources when multiple-user, open-access resources are converted into single-user, single-owned ones (Barbier and

Sathirathai 2004). Shrimp-pond operations themselves may affect the environment because they regularly use large quantities of antibiotics to protect the stock from infections (Benbrook 2002; FAO 2002). Producing shrimp using intensive methods is a risky and volatile activity because shrimp are vulnerable organisms that are more subject than most seas creatures to negative impacts from weather and diseases. Furthermore, shrimp feed contains a high proportion of fishmeal, which increases the already strong pressure on oceanic fisheries (the so-called "fish trap"), thereby undermining the sustainability of fish farming (Brown Larsen, and Fischlowitz-Roberts 2003; Naylor et al. 2000). Handing out this feed may itself result in eutrophication of the surface water and pollution of the pond bottom. Finally, the discharge of pond water and pond-bottom sludge may impact surrounding ecosystems through soil and surface-water salinization and by spreading pesticides, bleaches, and antibiotics.

A number of societal actors are developing interventions to address these serious environmental problems. It is interesting to witness a shift from government-dominated regulatory approaches to ones in which non-governmental actors are becoming increasingly important. This includes a growing involvement of foreign actors in addition to Thai agencies. Thus, Thai governmental regulations established in the early 1990s have become supplemented over time by private regulatory arrangements, such as certification and labelling schemes and consumer boycotts.

The Thai Department of Fisheries was tightening its environmental regulations as early as 1991 (Kaosa-ard and Wijukprasert 2000). Unfortunately, these measures were largely ineffective because most shrimp farmers did not comply with them; also, the authority to monitor and enforce compliance remained fragmented among several agencies (Flaherty, Vandergeest, and Miller 1999).[5] Efforts to create stricter legislation and to enforce existing laws have been blocked repeatedly by protests from the industry (Lebel et al. 2002).

Government-based regulations have largely failed because production has changed locations swiftly to stay ahead of them and also because of how the trade in shrimp is structured — that is, the sites of production and the sites of consumption are often in different jurisdictions. Stated another way, many environmental problems are geographically located beyond the authority of individual nation-states. At the same time, international regulations have a limited capacity to address environmental problems of this sort. The WTO agreements do allow national governments to intervene in the shrimp trade to ensure that specific food-safety standards are met, but such agreements are not intended to prescribe methods of shrimp production, processing, and trade (OECD 1994). The result is that environmental governance of global flows of shrimp increasingly involves non-governmental actors. Recently, for example, private firms have introduced codes of conduct and labelling schemes (most of the latter in collaboration with NGOs).

Different codes of conduct have been developed over the past decade, often inspired by the FAO Code of Conduct for Responsible Fisheries (NACA 2005). The FAO Code of Conduct, developed in 1994–95, contains specific guidelines to avoid environmental risks and to ensure the safety of the food produced. To achieve these goals, the FAO argues, national governments should create legal frameworks for place-based governance and establish international co-operation to govern the transboundary effects of aquacultural practices. The implementation of this code remains, however, rather restricted (Read and Fernandes 2003; Tacon and Forster 2003), because developing countries have limited institutional capacities and find it difficult to accommodate such place-bounded regulatory requirements with the need for regulation in the space of flows. The Holmenkollen guidelines for sustainable development, formulated in 1994 by a group of scientists, authorities, producers, and NGOs and adopted in 1998, are based on the conviction that modern aquaculture could become an important provider of food, provided that sustainable-development principles are applied — notably, the precautionary principle and the principle of human equity. Technological transitions toward "integrated, poly-culture-based fish farming for omnivorous or herbivorous species" are considered realistic solutions because ultimately all of those concerned have a common interest in clean water, a healthy environment, abundant wild stocks, and economic prosperity. A dialogue among all parties using sound, objective scientific knowledge should result in regulatory standards that can be legally enforced (Goode and Whoriskey 2003). The Code of Practice for Responsible Aquaculture, developed by the Global Aquaculture Alliance (GAA 2001), an international industry-based aquaculture organization, is developing a framework for environmentally and socially responsible shrimp farming. This code includes provisions for on-site inspections and controls; however, it does not yet require certification and labelling of seafood at the retail level (Roheim 2003).[6]

Global shrimp supply connects shrimp ponds in Thailand with the dinner tables in Europe, United States, and elsewhere. The environmental labelling of shrimp enables consumers to ensure the environmental quality and safety of the product. Certification schemes commonly used for health and safety in food production and processing, such as ISO 14001, HACCP (Hazard Analysis Critical Control Point), are also applied in shrimp aquaculture. ISO 14001 requires firms to establish an environmental policy and to set targets and objectives for their environmental-management performance (Frankic and Hershner 2003). ISO 14001 also allows a "green chain" certification from production to disposal. HACCP amounts to a preventive human-health risk-management system and is especially relevant for processed seafood. In 1997 the Thai government introduced the Thai Code of Conduct for Sustainable Shrimp Farming. That code was meant to ensure that the industry would be environmentally, socially, and economically responsible (Nissapa et al. 2002). An actor in the shrimp-production chain that abides by this code can acquire certification, and the shrimp it produces can be labelled "Thai Quality Shrimp" (Oosterveer 2006). Labelling of shrimp on the basis of such certification schemes creates new relationships between the shrimp producer and the consumers; it also links local shrimp-production practices with the places of shrimp consumption, via the global space of food flows.

Boycotts of farmed shrimp take consumer involvement even further. International NGOs such as the Environmental Justice Foundation, Greenpeace, and Christian Aid (Maybin and Bundell 1996; Barnhizer 2001; EJF 2003) link the environmental consequences of shrimp farming in developing countries with devastating impacts on local populations, whose livelihoods are destroyed. They consider the proposed technological improvements to be inadequate, because they would mainly solve the shrimp industry's problems and not those of the local populations. They argue that Western consumers should boycott shrimp produced through these "anti-social and environmentally destructive" practices and promote a transition toward extensive shrimp-farming practices controlled by local people combined with the capture of wild shrimps through sustainable methods.

These examples indicate how much variety is opened up by the innovative governance arrangements that are emerging; they also point to the growing involvement of various social actors in responses to the environmental and social impacts of shrimp farming.[7] The global character of shrimp production, trade, and consumption and the great differences in production practices have turned environmental governance of shrimp farming into a complicated exercise. The environmental problems resulting from aquaculture are still mostly localized; at the same time, the actual locations are constantly shifting as a result of ecosystem destruction and economic opportunities. Shrimp production and consumption has been organized into a permanent and increasingly complex global flow, one that involves producers, processors, traders, and retailers along with many others. The conventional government tools, based as they are on a clear separation between governmental and private responsibilities and between national and multilateral regulations, are no longer adequate. In many of the newly introduced governance arrangements, these divisions are no longer relevant. Consumer boycotts have a global character, whereas these arrangements are attempting to relocalize the governance of shrimp farming. Producer guidelines, certification schemes, and codes of practice provide incentives to producers to supplement national governmental regulations and thereby improve their environmental performance. Consumer-oriented arrangements such as ecolabels fit the present structure of global shrimp supply but they also recreate linkages between the space of places and the space of flows. They involve consumers as key actors in the global food networks alongside governments, producers, and traders/retailers.[8]

The sociology of environmental flows understands shrimp production, trade, and consumption, including their environmental impacts, as a material and informational flow. By combining this flow analysis with a study of the various societal actors involved and located in different, sometimes distant, places (or by analyzing the different scapes), it becomes possible to better understand why new arrangements emerge in the environmental governance of global shrimp supply. This approach clarifies in particular the interaction between global dynamics and local impacts.

Invasive Species: Governing Biodiversity Flows in Global Spaces

Biodiversity, too, can be conceptualized as a global environmental flow, though one that functions in a completely different way than global shrimp supply. Biodiversity is a material flow in terms of the migration of specific animals and the dispersion of plants, micro-organisms, and genetic material; it also includes the informational and financial flows associated with these travelling organisms and genes. The following section discusses the concept of global biodiversity flows and then reviews the particular global biodiversity flow of "invasive species," the (potential) environmental impacts of that flow, and some of the governance arrangements suggested to address those impacts.

Initially, biodiversity governance was perceived as the science-based preservation of genetic diversity per se, replacing the "classical" practice of nature conservation, which was founded on the preservation of aesthetically and emotionally appreciated wildlife and wilderness. Nowadays, controlling specific transnational material flows (i.e., the migration of specific animals and the dispersion of certain plants and micro-organisms) linked to flows of data and expertise, money, concepts, and regulatory measures is considered vital to the maintainance of biodiversity in conditions of global change.[9] Global biodiversity flows determine more and more which values are attributed to nature, which exotic species are to be refused, which areas get priority in conservation, which drivers and consequences are to be included in projects, who is to be involved in monitoring, and which discourses and which actors will govern decisions on nature protection and construction (Van Koppen 2006). Clearly, biodiversity management should no longer be understood simply as the protection of as many different species as possible, but as the introduction of innovative global arrangements to govern global environmental flows through networks and scapes involving both governmental and non-governmental actors.

A particular global biodiversity problem concerns "invasive species," whose governance is currently closely associated with the intensifying linkages between global and local dynamics in ecosystems, with trade and transport. Invasive species constitute a global environmental flow involving continuous local, transnational, and even global movements of animals, plants, fungi, genes, and so on. This material flow is closely related to informational and conceptual flows. Though the movement of species has always occurred in natural history, the increase in global trade is viewed as contributing to the growing threat that "non-indigenous" species pose to existing ecosystems and to biodiversity in general. Within the global space of flows, people–biodiversity interactions may take place across large distances in space and time. Organisms may travel "on purpose" (i.e., when people transport certain plants or animals); or they may travel unintentionally as part of other flows, such as those of agricultural crops, livestock, fish, medicines, cultivated and wild horticultural plants, or plant oils. A particular risk seems to be the ballast water that ships take on board when they travel without cargo: this water may contain various aquatic organisms, which then spread through non-native ecosystems when it is released at a distant harbour. In this way many different species, such as insects, fungi, and pathogens (and even genes, in the case of GM-food production and trade) cross (ecosystem) borders unintentionally.

Invasive or exotic species pose a potential environmental risk because an organism that constitutes an unproblematic part of biodiversity at one place may become a danger to biodiversity at another location, thus creating a tension between the global space of flows and the local space of place. This phenomenon is often seen as a major and irreversible threat to environmental sustainability (Campbell 2001; Burgiel et al. 2006). Invasive species are considered a threat to the survival of endangered species; they are also seen as affecting ecosystem processes, reducing crop yields, infecting livestock, and endangering human health. Recently, Pimentel and colleagues (2005) estimated the annual damage and loss caused by alien-invasive species in the United States alone at US$120 billion. However, scientific knowledge of biodiversity is still limited, and therefore so is knowledge about the (potential) impacts of certain invasive species on particular ecosystems.[10] Some observers even contend that non-native species can be beneficial to food production or biodiversity; for example, introduced species provide more than 98 percent of the food produced by the American food system. Clark (2002) states that nature, or ecosystems, should not be understood as having a balance that can be upset by exotics because nature at the level of biotic assemblages has never been homeostatic. Most "climax" ecosystems are a fairly heterogeneous mix of relatively mature and more recently perturbed patches (ibid.). Humans have always contributed, intentionally or not, to modifications in the surrounding nature. So it is essential "to recognize that disturbance, like mobilism, invasion and hybridisation, is endemic to the living world" (ibid., 114). And even native species can lead to costly or disruptive consequences for local biodiversity. In other words, both native and non-native species can be harmful, beneficial, or both, and it is often impossible to determine which beforehand. The complexity of biodiversity flows means that not every intervention is noxious and that not all non-linear events are necessarily catastrophic.

It is a complex and challenging task to govern invasive species, for living biodiversity material is volatile. The scientific debate continues regarding the environmental impacts of that material. Nevertheless, over the past thirty years various initiatives have been taken to govern this global environmental flow. Initially the focus was on governmental intervention, and several multilateral conventions have been formulated to control particular intentional flows of global biodiversity. An example is the CITES convention to prevent uncontrolled trade in endangered species.[11] CITES and other international agreements are essentially based on border control of international biodiversity flows so that the harmless can be separated from the harmful. It has often been suggested that this approach is the most adequate one for ensuring the same level of environmental protection as existed in the past (Burgiel et al. 2006). However, monitoring environmental risks through national authorities — that is, inspecting for potential invasive species at customs posts and blocking them when found — seems prohibitively costly. Also, strict border control is incompatible with the objective of facilitating international trade as laid down by the WTO, to which most countries belong. According to the WTO's TBT and SPS agreements, trade-control measures are only acceptable when justified by clear scientific evidence. The currently existing scientific uncertainty about the immediate and long-term effects of specific transnationally travelling species makes fulfilling this requirement a challenge.

Besides governments, many non-governmental actors are increasingly involved in governing the global flows of invasive species. The public debate on the dangers of invasive species to global biodiversity is taking place mainly within scientific and environmental-NGO circles. As a result of these debates, national governments have come under pressure to establish multilateral agreements, such as the Cartagena Protocol of the Convention on Biodiversity (CBD). Implementing the resulting measures necessarily involves many different social actors engaged in international trade and transport. Not all actors concerned, however, have similar interests: some intend to control and limit transnational movements as much as possible to reduce possible risks; others intend to *facilitate* such flows.[12] The case of the public debates surrounding the international trade in genetically modified organisms, for example, involved farmers' and consumers' organizations, biotech companies, supermarkets, scientists, and environmental NGOs; each had particular interests that led to a variety of alliances (Oosterveer 2005).

An especially challenging problem, then, is how to identify the relevant social actors in the social networks related to the governance of global flows of invasive species. In contrast to the rather clearly definable global shrimp-supply chain, it is often difficult to establish who is involved in biodiversity flows. In particular, it is a challenge to ascertain the relevant social actors among the wide variety involved in the unintended global movements of organisms closely related to other global material flows. In many situations these relevant social actors do not constitute a stable and clearly visible network. A further complexity is that decisions regarding which material flows should be considered invasive are the result of ongoing scientific, societal, and political debates. The recent discussions on the spread of bird flu in Asia and Europe, generally associated with migratory birds and the international poultry trade, make it very clear how invasive species can unexpectedly involve groups of social actors previously not concerned.

Compared to many other environmental flows, global flows of invasive species are highly fluid. Their management is more challenging because they can have dramatic impacts despite great uncertainty. Biodiversity in general, and invasive species in particular, are contested concepts. Governing global flows of invasive species requires global information systems to establish which organisms may be dangerous in which particular locations. Also, global flows of invasive species are exceptional in the sense that—more than is typical—the material flows and the related conceptual/informational flows have their own internal dynamics. Nevertheless, the material and social dimensions of flows of invasive species remain intimately intertwined and can be considered a global environmental flow.

Conceptualizing invasive species as a global material flow from the perspective of the sociology of environmental flows suggests an analytical focus on the linkages among the material, social, and conceptual/informational dimensions of this flow. This perspective clarifies in particular the interaction between global and local dynamics, the environmental impacts from social activities at a distance, the fluidity of material flows, and the ways in which societal and scientific debates on concepts and definitions have real material effects. The particular challenges facing the governance of global flows of invasive species become evident in the difficulties and uncertainties surrounding the process of identifying and defining the flow itself, establishing who the relevant social actors are, and determining the applicability of different governance options. Because it can combine a material-flows analysis with a study of the various societal actors involved in different places, the sociology-of-flows perspective offers indications for potential innovative governance arrangements that better fit the hybrid character of global flows of invasive species.

Conclusion

This chapter has assessed whether a sociology-of-flows approach might be helpful in analyzing the contemporary governance of environmental problems in technonatural time/spaces. Judging from the cases discussed, this approach offers interesting opportunities to better understand current dynamics in environmental governance.

The transition toward a global-network society underscores the growing worldwide interconnectedness of environmental problems, the need to constantly link local and global dynamics, and the intimate links between social and material dimensions. The flows concept points to the evolving interactions among various social actors—located at different places, active at varying scales, connected to movements of particular materials, governmental and non-governmental—in the definition and governance of environmental problems. Understanding the roles of these social actors and their interactions in relation to environmental problems requires that we avoid predetermined definitions of the material issues at hand. In environmental problems, material and social dimensions cannot possibly be disentangled in advance, and neither can the spatial dimensions of "the local" and "the global." Environmental problems are not simply the (external) material impacts of social activities, such as producing or consuming goods, transport, and waste. Local and global dynamics interact permanently, and as a result of varying material and social dimensions they create different scapes. The concept of environmental flows makes it possible to connect these dimensions in a more integrated manner. This concept also points at the need to leave behind some conventional views on environmental governance. With the help of a "sociology of environmental flows," different innovative, more flexible, and more adequate environmental governance arrangements can be identified. Nevertheless, improved understanding of such potential environmental governance arrangements demands further reflection on the dynamics in different dimensions, local/global and material/social, because increased flexibility does not necessarily signify the absence of any structure.

Notes

1 This chapter is based on collaborative work in the Environmental Policy Group at Wageningen University.

2 The "space of flows" still includes a territorial dimension because it requires a technological infrastructure that operates from certain locations connecting functions and people located in specific places. But these network nodes are less oriented toward the specific geographical characteristics of the location and its surroundings and much more toward interactions with the other nodes in the network.

3 "Open systems" shrimp farming near the sea requires high volumes of saline water, exchanging between 30 and 40 percent of pond volume per day during the later growth stages to maintain the water quality. Despite the shift to "closed systems," the latter still need to add fresh and salt water throughout the grow-out phase to offset losses from water seepage and evaporation. Interestingly, the balance between fresh and salt water changed when low-salinity techniques were developed "through the efforts of innovative small-scale farmers" (Flaherty and Vandergeest 1998, 822).

4 Shrimp-processing industries in Thailand were employing some 60,000 workers during the mid-1990s (Thailand Department of Fisheries), and this labour is particularly intensive, where young women work for long hours for low pay (Goss and Rickson 2000).

5 In addition, authorities continued subsidizing shrimp-farm development and lifted the ban on logging mangrove areas.

6 The GAA claims that there are no particular risks involved in shrimp farming because when properly conducted it is a profitable, environmentally sound, and socially beneficial activity. As with any young and rapidly growing industry, mistakes are made. Even so, such negative environmental and social impacts "have invariably resulted from poor planning or poor management by shrimp farmers and government agencies rather than as a routine consequence of shrimp farming" (GAA 2001).

7 However, the concrete impacts of various non-governmental arrangements may well be more on the (organization of the) world's fishing industries and less on the fisheries themselves (Roheim and Sutinen, 2006).

8 So far no international guidelines exist for organic aquaculture production, though organic labelling of shrimp might contribute considerably to more sustainable use of aquatic resources and offer interesting marketing opportunities for small producers (Bailly and Willmann, 2001). However, the organic movement in the United States has denied the possibility of applying the same criteria on both terrestrial and aquatic animals (fish); therefore, farmed fish can never be considered organic (Mansfield 2003, 2004). Organic aquaculture standards were developed by members of the

International Federation of Organic Agriculture Movements (IFOAM) and finalized and included in the IFOAM Basic Standards by the General Assembly in 2005. By 2008, fourteen global organic standard-setting organizations had accepted these standards for the organic rearing of aquatic species. Note that the U.S. National Organics Standards Board was not among these organizations. The IFOAM reports a global area of 4,080 hectares of certified organic aquaculture in 2007.

9 Biodiversity scientists have a particularly strong position in the networks in that they frame the debate and often prescribe other social actors' behaviour (Gaston 1996; Swanson 1997), though some biodiversity researchers openly acknowledge the scientific uncertainties surrounding the ecosystem functions of biodiversity (Kunin and Lawton 1996; McCann 2000; Perrings et al. 1995; UNEP 1995).

10 For example, probably less than 5 percent of the earth's fungi have been named by science (Campbell 2001).

11 The main global conventions that are biodiversity-related (Pritchard, 2005:
  – the Convention on Biodiversity (CBD), 1992
  – the Convention on Wetlands (the Ramsar Convention), 1971
  – the Convention on Migratory Species (CMS, or Bonn Convention), 1979
  – the Convention on International Trade in Endangered Species (CITES), 1973
  – the Convention for the Protection of the World Cultural and Natural Heritage (the World Heritage Convention), 1972

12. This discussion becomes even more complicated when mad-made biodiversity (resulting from gen-technology) is also taken into account.

Works Cited

Adam, B. 2000. "The Temporal Gaze: The Challenge for Social Theory in the Context of GM Food." *British Journal of Sociology* 51: 125–42.

Appadurai, A. 1996. *Modernity at Large: Cultural Dimensions of Globalization.* Minneapolis: University of Minnesota Press.

Ayres, R.U., L.W. Ayres, P. Frankl, H. Lee, and O.M. Weaver. 1996. *Industrial Ecology: Towards Closing the Materials Cycle.* London: Edward Elgar.

Bailly, D., and R. Willmann. 2001. "Promoting Sustainable Aquaculture through Economic and Other Incentives." In *Aquaculture in the Third Millennium*, ed. R.P. Subasinhe, P. Bueno, M.J. Phillips, C. Hough, S.E. McGladdery, and J.R. Arthur. Bangkok: NACA and FAO. 95–101.

Barbier, E.B., and S. Sathirathai, eds. 2004. *Shrimp Farming and Mangrove Loss in Thailand.* Cheltenham: Edward Elgar.

Barnhizer, D., ed. 2001. *Effective Strategies for Protecting Human Rights.* Aldershot: Ashgate.

Bauman, Z. 1987. *Legislators and Interpreters*. Cambridge: Polity.

Beardsworth, A., and T. Keil. 1997. *Sociology on the Menu: An Invitation to the Study of Food and Society*. London and New York: Routledge.

Beck, U. 1997. *Was ist Globalisierung? Irrtümer des Globalismus — Antworten auf Globalisierung*. Frankfurt am Main: Suhrkamp.

Beck, U., and N. Sznaider. 2006. "Unpacking Cosmopolitanism for the Social Sciences: A Research Agenda." *British Journal of Sociology* 57: 1–23.

Benbrook, C. 2002. "Antibiotic Drug Use in US Aquaculture." Minneapolis: Institute for Agriculture and Trade Policy (IATP).

Brown, L.R., J. Larsen, and B. Fischlowitz-Roberts. 2003. *The Earth Policy Reader*. London: Earth Policy Institute and Earthscan.

Bunker, S. 1996. "Raw Material and the Global Economy: Oversights and Distortions in Industrial Ecology." *Society and Natural Resources* 9: 419–29.

Burgiel, S., G. Foote, M. Orellana, and A. Perrault. 2006. "Invasive Alien Species and Trade: Integrating Prevention Measures and International Trade Rules." Washington: Center for International Environmental Law (CIEL) and Defenders for Wildlife.

Buttel, F.H., G. Spaargaren, and A.P.J. Mol. 2006. "Epilogue: Environmental Flows and Early Twenty-First Century Environmental Social Sciences." In *Governing Environmental Flows in Global Modernity*, ed. F.H. Buttel, G. Spaargaren, and A.P.J. Mol. Cambridge, MA: MIT Press. 351–69.

Campbell, F.T. 2001. "The Science of Risk Assessment for Phytosanitary Regulation and the Impact for Changing Trade Regulations." *BioScience* 52: 148–53.

Castells, M. 2004. "Informationalism, Networks, and the Network Society: A Theoretical Blueprint." In *The Network Society: A Cross-Cultural Perspective*, ed. M. Castells. Cheltenham: Edward Elgar. 3–45.

———. 1998. *End of Millennium*. Vol. III of *The Information Age: Economy, Society, and Culture*. Oxford: Blackwell.

———. 1997. *The Power of Identity*. Vol. II of *The Information Age: Economy, Society, and Culture*. Oxford: Blackwell.

———. 1996. *The Rise of the Network Society*. Vol. I of *The Information Age: Economy, Society, and Culture*. Oxford: Blackwell.

Catton, W.R., and R.E. Dunlap. 1978. "Environmental Sociology: A New Paradigm." *American Sociologist* 13: 41–49.

Clark, N. 2002. "The Demon-Seed: Bioinvasion as the Unsettling of Environmental Cosmopolitanism." *Theory, Culture, and Society* 19: 101–25.

Dicken, P., P.F. Kelly, K. Olds, and H. Wai-Chung Yeung. 2001. "Chains and Networks, Territories and Scales: Towards a Relational Framework for Analysing the Global Economy." *Global Networks* 1: 89–112.

Duraiappah, A., A. Israngkura, O. Kuik, S. Hare, S. Pendekar, and D. Patmasiriwat. 2000. "Sustainable Shrimp Farming in Thailand." In *CREED-Policy Brief*. Amsterdam and Bangkok: IVM and TDRI.

EJF (Environmental Justice Foundation). 2003. "Squandering the Seas: How Shrimp Trawling Is Threatening Ecological Integrity and Food Security around the World." London: Environmental Justice Foundation.

FAO (Food and Agriculture Organization). 2002. "The State of World Fisheries and Aquaculture: 2002." Rome.

Fine, B. 1998. *The Political Economy of Diet, Health, and Food Policy*. New York and London: Routledge.

Flaherty, M., and P. Vandergeest. 1998. "'Low-salt' Shrimp Aquaculture in Thailand: Goodbye Coastline, Hello Khon Kaen!" *Environmental Management* 22: 817–30.

Flaherty, M., P. Vandergeest, and P. Miller. 1999. "Rice Paddy or Shrimp Pond: Tough Decisions in Rural Thailand." *World Development* 27: 2045–60.

Folke, C., N. Kautsky, H. Berg, A. Jansson, and M. Troell. 1998. "The Ecological Footprint Concept for Sustainable Seafood Production: A Review." *Ecological Applications* 8: S63–S71.

Frankic, A., and C. Hershner. 2003. "Sustainable Aquaculture: Developing the Promise of Aquaculture." *Aquaculture International* 11: 517–30.

GAA (Global Aquaculture Alliance). 2001. "Codes of Practice for Responsible Shrimp Farming." St. Louis.

Gaston, K., ed. 1996. *Biodiversity: A Biology of Numbers and Differences*. Oxford: Blackwell.

Giddens, A. 1979. *Central Problems in Social Theory: Action, Structure, and Contradiction in Social Analysis*. Berkeley: University of California Press.

Goldberg, R., M. Elliott, and R. Naylor. 2001. "Marine Aquaculture in the United States: Environmental Problems and Policy Options." Arlington: PEW-Oceans Commission.

Goldfrank, W.L., David Goodman, and A. Szasz, Eds. 1999. *Ecology and the World-System*. Westport, CT: Greenwood.

Goode, A., and F. Whoriskey. 2003. "Finding Resolution to Farmed Salmon Issues in Eastern North America." In *Salmon at the Edge*, ed. Derek Mills. Oxford: Blackwell. 144–58.

Goodman, D. 2001. "Ontology Matters: The Relational Materiality of Nature and Agro-Food Studies." *Sociologia Ruralis* 41: 182–200.

Goodman, D., and M. Watts, ed. 1997. *Globalising Food: Agrarian Questions and Global Restructuring*. London: Routledge.

Goss, J., D. Burch, and R. Rickson. 2000. "Agri-Food Restructuring and Third World Transnationals: Thailand, the CP Group, and the Global Shrimp Industry." *World Development* 28: 513–30.

Graedel, T.E., and B.R. Allenby. 1995. *Industrial Ecology*. Englewood Cliffs, NJ: Prentice Hall.

Gräslund, S., and B. Bengtsson. 2001. "Chemicals and Biological Products Used in South-East Asian Shrimp Farming, and Their Potential Impact on the Environment—A Review." *Science of the Total Environment* 280: 93–131.

Held, D., A. McGrew, D. Goldblatt, and J. Perraton. 1999. *Global Transformations: Politics, Economics, and Culture.* Cambridge: Polity.

Huitric, M., C. Folke, and N. Kautsky. 2002. "Development and Government Policies of Shrimp Farming Industry in Thailand in Relation to Mangrove Ecosystems." *Ecological Economics* 40: 441–55.

Josupeit, H. 2004. "An Overview on the World Shrimp Market." In *Globefish: World Shrimp Markets.* Madrid: Globefish.

Kaosa-ard, M., and P. Wijukprasert. 2000. "The State of Environment in Thailand: A Decade of Change." Bangkok: Thailand Development Research Institute.

Karkkainen, B. 2004. "Post-Sovereign Environmental Governance." *Global Environmental Politics* 4: 72–96.

Keohane, R., and J. Nye. 1977. *Power and Independence: World Politics in Transition.* Boston: Little, Brown.

Krasner, S., ed. 1983. *International Regimes.* Ithaca, NY: Cornell University Press.

Kunin, W., and J. Lawton. 1996. "Does Biodiversity Matter? Evaluating the Case for Conserving Species." In *Biodiversity: A Biology of Numbers and Differences,* ed. K. Gaston. Oxford: Blackwell. 283–308.

Lebel, L., N. Tri, S. Saengnoree, S. Pasong, U. Buatama, and L. Thoa. 2002. "Industrial Transformation and Shrimp Aquaculture in Thailand and Vietnam: Pathways to Ecological, Social, and Economic Sustainability." *Ambio* 31: 311–23.

Lipschutz, R., and K. Conca, eds. 1993. *The State and Social Power in Global Environmental Politics.* New York: Columbia University Press.

Lipschutz, R., and C. Fogel. 2002. "'Regulation for the Rest of Us'? Global Civil Society and the Privatization of Transnational Regulation." In *The Emergence of Private Authority in Global Governance,* ed. R. Hall and T. Biersteker. Cambridge: Cambridge University Press. 115–40.

Litfin, K., ed. 1998. *The Greening of Sovereignty in World Politics.* Cambridge, MA: MIT Press.

Mansfield, B. 2004. "Organic Views of Nature: The Debate over Organic Certification for Aquatic Animals." *Sociologia Ruralis* 44: 216–32.

———. 2003. "From Catfish to Organic Fish: Making Distinctions about Nature as Cultural Economic Practice." *Geoforum* 34: 329–42.

Maybin, E., and K. Bundell. 1996. "After the Prawn Rush. The Human and Environmental costs of Commercial Prawn Farming." London: Christian Aid.

McCann, K. 2000. "The Diversity-Stability Debate." *Nature* 405: 228–33.

Mol, A.P.J. 2001. *Globalization and Environmental Reform. The Ecological Modernization of the Global Economy.* Cambridge, MA: MIT Press.

Mol, A.P.J., and Fred H. Buttel, eds. 2002. *The Environmental State under Pressure.* London: Elsevier/JAI.

Mol, A.P.J., and G. Spaargaren. 2002. "Ecological Modernization and the Environmental State." In *The Environmental State under Pressure*, ed. F.H. Buttel and A.P.J. Mol. Oxford: Elsevier Science. 33–52.

Mol, A.P.J., and D.A. Sonnenfeld. 2000. "Ecological Modernisation around the World: An Introduction." In *Ecological Modernisation around the World: Perspectives and Critical Debates*, ed. A.P.J. Mol and D.A. Sonnenfeld. London: Frank Cass. 3–14.

Murdoch, J. 2000. "Networks — A New Paradigm of Rural Development?" *Journal of Rural Studies* 20: 407–19.

NACA (Network of Aquacultural Centres in Asia-Pacific). 2005. "Certification Schemes."

Naylor, R.L., R. Goldburg, J. Primavera, N. Kautsky, M. Beveridge, J. Clay, C. Folke, J. Lubchenko, H. Mooney, and M. Torrell. 2000. "Effect of Aquaculture on World Fish Supplies." *Nature* 405: 1017–24.

Nissapa, A., S. Boromthanarat, B. Chaijaroenwatana, and W. Chareonkunanond. 2002. "Shrimp Farming in Thailand: A Review of Issues." EU INCO-Dev Project PORESSFA No. IC4-CT-2001-10042.

OECD (Organisation for Economic Co-operation and Development). 1994. "Trade and Environment: Processes and Production Methods." Paris.

Oosterveer, P. 2007. *Global Governance of Food Production and Consumption: Issues and Challenges.* Cheltenham: Edward Elgar.

———. 2006. "Globalization and Sustainable Consumption of Shrimp: Consumers and Governance in the Global Space of Flows." *International Journal of Consumer Studies* 30: 465–76.

———. 2005. "Global Food Governance." Wageningen: Wageningen University.

Perrings, C., K.G. Maler, C. Folke, C.S. Holling, and B.O. Jansson, eds. 1995. *Biodiversity Loss: Economic and Ecological Issues.* Cambridge: Cambridge University Press.

Pillay, T.V.R. 1992. *Aquaculture and the Environment.* Oxford: Fishing News.

Pimentel, D., R. Zuniga, and D. Morrison. 2005. "Update on the Environmental and Economic Costs Associated with Alien-Invasive Species in the United States." *Ecological Economics* 52: 273–88.

Pritchard, D. 2005. "Biodiversity-Related Treaties: International Biodiversity-Related Treaties and Impact Assessment — How Can They Help Each Other?" *Impact Assessment and Project Appraisal* 23: 7–16.

Read, P., and T. Fernandes. 2003. "Management of Environmental Impacts of Marine Aquaculture in Europe." *Aquaculture* 226: 139–63.

Roberts, J., and P. Grimes. 2003. "World-System Theory and the Environment: Toward a New Synthesis." In *Sociological Theory and the Environment*, ed.

R. Dunlap, F.H. Buttel, P. Dickens, and A. Gijswijt. Lanham: Rowman and Littlefield. 167–94.

Roheim, C. 2003. "The Seafood Consumer: Trade and the Environment." In *The International Seafood Trade*, ed. James L. Anderson. Boca Raton and Cambridge: CRC Press and Woodhead. 193–204.

Roheim, C., and J.G. Sutinen. 2006. "Trade and Marketplace Measures to Promote Sustainable Fishing Practices." In *ICTSD Natural Resources, International Trade, and Sustainable Development*. Geneva: ICTSD (International Centre for Trade and Sustainable Development) and the High Sea Task Force.

Schnaiberg, A. 1980. *The Environment: From Surplus to Scarcity*. Oxford: Oxford University Press.

Shaw, M. 2000. *Theory of the Global State: Globality as an Unfinished Revolution*. Cambridge: Cambridge University Press.

Spaargaren, G., A.P.J. Mol, and F.H. Buttel, eds. 2006a. *Governing Environmental Flows: Global Challenges to Social Theory*. Cambridge, MA: MIT Press.

———. 2006b. "Introduction: Governing Environmental Flows in Global Modernity." In *Governing Environmental Flows: Global Challenges to Social Theory*, ed. G. Spaargaren, A.P.S. Mol, and F.H. Buttel. Cambridge, MA: MIT Press. 1–36.

Spangenberg, J.J., A. Femia, F. Hinterberger, and H. Schütz. 1998. "Material Flow–Based Indicators in Environmental Reporting." In *Environmental Issues Series*, no. 14. Copenhagen: European Environmental Agency. 58.

Stevis, D., and A. Assetto, eds. 2001. *The International Political Economy of the Environment*. Boulder, CO: Lynne Riener.

Stevis, D., and H. Bruyninckx. 2006. "Looking through the State at Environmental Flows and Governance." In *Governing Environmental Flows: Global Challenges to Social Theory*, ed. G. Spaargaren, A.P.J. Mol, and F.H. Buttel. Cambridge, MA: MIT Press. 107–36.

Swanson, T. 1997. *Global Action for Biodiversity*. London: Earthscan.

Tacon, A., and I. Forster. 2003. "Aquafeeds and the Environment: Policy Implications." *Aquaculture* 226: 181–89.

UNEP (UN Environmental Programme). 1995. *Global Biodiversity Assessment*. Cambridge: Cambridge University Press.

Urry, J. 2003. *Global Complexity*. Cambridge: Polity.

———. 2000. *Sociology beyond Society*. London: Routledge.

Van der Meulen, H. 2000. *Circuits in de Landbouwvoedselketen. Verscheidenheid en Samenhang in de Productie en Vermarkting van Rundvlees in Midden-Italië*. Wageningen: Circle for Rural European Studies.

Van Koppen, C.S.A. (Kris). 2006. "Governing Nature? On the Global Complexity of Biodiversity Conservation." In *Governing Environmental Flows: Global Challenges to Social Theory*, ed. G. Spaargaren, A.P.J. Mol, and F.H. Buttel. Cambridge, MA: MIT Press.

Wapner, P. 1996. *Environmental Activism and World Civic Politics*. Albany, NY: SUNY Press.

Young, O. 2000. *Global Governance: Drawing Insights from Environmental Experience*. Cambridge: MIT Press.

———. 1994. *International Governance: Protecting the Environment in a Stateless Society*. Ithaca, NY: Cornell University Press.

Chapter Two

Circulations and Metabolisms: (Hybrid) Natures and (Cyborg) Cities
Erik Swyngedouw

Cyborg Cities: The Urbanization of Nature
Imagine standing on Piccadilly Circus in London and considering the socio-environmental metabolic relations that come together in this global–local place. Smells, tastes, things, and bodies from all nooks and crannies of the world are floating by, consumed, displayed, narrated, visualized, and transformed. The Amazon Forest Shop and Restaurant plays to the tune of ecosensitive shopping and the multibillion-pound eco-industry while competing with McDonald's burgers and Dunkin' Donuts. The sounds of world music vibrate from Virgin's Megastore, while people, spices, clothes, foodstuffs, and materials from all over the planet whirl by. The neon lights are fed by nuclear processes, or by coal or gas burning in far-off power plants, while passing cars consume fuels from oil deposits and pump CO_2 into the air, affecting forests, climates, and people around the globe. These disparate processes trace the global geographic mappings that flow through the urban and "produce" cities as palimpsests of densely layered bodily, local, national, and global — but geographically depressingly uneven — socio-ecological and technonatural processes. This intermingling of things material and things symbolic produces a particular socio-environmental milieu that welds nature, society, and the city together, often through many layers of networked technostructures (such as pipes, cables, relay stations, logistical apparatuses, and the like), into a deeply heterogeneous, conflicting, and often disturbing whole (Swyngedouw 1996).

In the summer of 1998 the Southeast Asian financial bubble burst. Global capital lurched from place to place, turning cities like Jakarta into social and physical wastelands where dozens of unfinished skyscrapers dotted the landscape while thousands of unemployed children, women, and men roamed the streets in search of survival. In the meantime, El Niño was wreaking havoc by disturbing the world's climate. In Jakarta, puddles of stagnant water formed around the abandoned skyscraper projects that had once promised continuing capital accumulation for Indonesia; these became breeding grounds and superb ecological niches for mosquitoes. Malaria and dengue fever suddenly joined with unemployment and social and political mayhem in shaping that cityscape. Global capital and the technoscapes of the world's financial architecture had fused with global climate, with local power struggles, and with socio-ecological conditions to reshape its social ecology in profound, radical, and deeply troubling ways.

Or consider this: about 750 million urban dwellers worldwide do not have access to clean and potable water, while others have an unlimited supply of water. Water-borne diseases are the primary cause of premature death in the developing world. Access to nature in the city can be — indeed, often *is* — a matter of life or death (Swyngedouw 2004).

The above examples illustrate how the city — and urbanization more generally — can be viewed as a process of deterritorialization and reterritorialization through metabolic circulatory flows that are organized through social and physical conduits or networks of "metabolic vehicles" (Virilio 1986). These processes are infused by relations of power in which social actors strive to defend and create their own environments in a context of class, ethnic, racial, and/or gender struggles. Under capitalism, the commodity relation and the flow of money veil the multiple socio-ecological processes of domination/subordination and exploitation/repression, which feed the urbanization process and turn the city into a metabolic socio-environmental process that reaches out from the immediate environment to the remotest corners of the globe (Kaika and Swyngedouw 1999).

Currently more than half the world's people live in cities — a proportion that will rise to nearly two-thirds by 2050. So it is no exaggeration to say that socio-ecological processes and effects, as well as global ecological issues and problems, are driven by, organized through, and shaped by the processes of urbanization itself. As Timothy Luke (2003, 12) maintains, cities "leave very destructive environmental footprints as their inhabitants reach out into markets around the world for material inputs to survive, but the transactions of this new political ecology also are the root causes of global ecological decline." This chapter argues that to fully grasp these developments we need to consider how nature becomes urbanized through proliferating socio-metabolic processes.

The first objective of this chapter is to foreground "circulation" and "metabolism" as possible entry points for theorizing about and analyzing socio-natural things. "Metabolism" and "circulation" embody what modernity has been and will always be about—that is, a series of interconnected heterogeneous (human and non-human) and dynamic, but contested and contestable, processes of continuous quantitative and qualitative transformations that rearrange humans and non-humans in new and often unexpected ways. By emphasizing movement, circulation, change, and process, and by insisting on the socially mobilized "materiality" of life, this chapter posits that historical materialism has been among the first social theories to productively embrace and mobilize "metabolism" and "circulation" as entry points in undertaking "ontologies of the present that demand archaeologies of the future" (Jameson 2002, 215).

The second, related objective is to mobilize "metabolism" and "circulation" as socio-ecological processes that permit us to frame questions of the environment—in particular, of the *urban* environment—in ways that are radically political. I propose a framework for analysis for an urban political ecology that allows us to draw together the insights of historical-geographical materialism with the work of Haraway to view the modern city as a process that fuses the social and the natural together to produce a distinct "hybrid" or "Cyborg" urbanization (see Luke 1996; Swyngedouw 1996; Gandy 2005). Cyborg metaphors have long been used to understand human–machine relations, but they also have uses for urban studies, as Matthew Gandy puts it:

> The emphasis of the cyborg on the material interface between the body and the city is perhaps most strikingly manifested in the physical infrastructure that links the human body to vast technological networks. If we understand the cyborg to be a cybernetic creation, a hybrid of machine and organism, then urban infrastructures can be conceptualized as series of interconnecting life support systems. The modern home, for example, has become a complex exoskeleton for the human body with a provision of water, warmth, light and

other essential needs. The home can be conceived as a "prosthesis and pro-phylactic" in which modernist distinctions between nature and culture, and between the organic and the inorganic, become blurred. (ibid., 28)

The chapter concludes by considering how developing new urban political ecologies has the potential to open up the theoretical and practical possibility of creating the (urban) environments we wish to inhabit. The urbanization of nature, though generally portrayed as a technological/engineering problem, is in fact as much part of the politics of life as any other social process. Recognition of this *political* meaning of nature is essential if sustainability is to be combined with just and empowering urban development—that is, with urban development that returns the city and the city's environment to its citizens.

Entering Metabolism and Circulation

The terms "metabolism" and "circulation" have a long history. The concept of "metabolism" arose in the early nineteenth century, mainly in relation to material exchanges in the body with respect to respiration. It later became extended to include material exchanges between organisms and the environment as well as the biophysical processes within living and non-living (i.e., decaying) entities.

For example, Jacob Moleschott (1857) and Justus von Liebig (1840, 1842) delineated the exchange of energy and substances between organisms and the environment on the one hand and the totality of biochemical reactions in a living thing on the other. Indeed, von Liebig viewed organisms as living processes, assigning them a history-as-process. Interestingly enough, von Liebig took the temporal/spatial separation of spaces of production and spaces of consumption through the emergence of long-distance trade on the one hand and the process of urbanization on the other (what he called the "metabolic rift") as pivotal causes affecting (negatively) the productivity of agricultural land on the one hand and the problematic accumulation of excrement, sewage, and garbage in the city on the other (Page 2004). Karl Marx would later incorporate this view in his historical materialism.

Marx and Engels were among the first to mobilize the term "metabolism" to grapple with the dynamics of socio-environmental change and evolution (Fisher-Kowalski 1998, 2003). In fact, "metabolism" is the central metaphor in Marx's definition of labour and in his analysis of the relationship between human and nature: "Labour is, first of all, a process between man and nature, a process by which man, through his own actions, mediates, regulates, and controls the *metabolism* between himself and nature...Through this movement he acts upon external nature and changes it, and in this way he simultaneously changes his own nature" (1970, 283, 290).

Indeed, for Marx, this socio-natural metabolism is the very foundation of history—that is, a socio-environmental history through which the natures of humans and non-humans alike are transformed (see also Godelier 1986). To the extent that labour constitutes the universal premise for human metabolic interaction with nature, the particular socio-technical vehicles and social relations through which this metabolism of nature is enacted shape the very form of labour. For historical materialism, then, ecology is not so much a question of values, morals, or ethics, but rather a means of "understanding the evolving material interrelations (what Marx called 'metabolic relations') between human beings and nature."

From a consistent materialist standpoint, "the question is...one of coevolution" (see also Norgaard 1994). Foster goes on to argue that

> a thoroughgoing ecological analysis requires a standpoint that is both materialist and dialectical...A materialist sees evolution as an open-ended process of natural history, governed by contingency, but open to rational explanation. A materialist viewpoint that is also dialectical in nature (that is, a non mechanistic materialism) sees this as a process of transmutation of forms in a context of interrelatedness that excludes all absolute distinctions...A dialectical approach forces us to recognize that organisms in general do not simply adapt to their environment; they also affect that environment in various ways by affecting change in it. (2000, 15–16; see also Levins and Lewontin 1985)

Marx undoubtedly borrowed the notion of "metabolic interaction" from von Liebig. In contrast to other sociologists such as Comte and Spencer, who used the concept of metabolism as an analogy for grappling with social metabolism and for whom "nature offered the gnoseological structures to survey the workings of society" (Padovan 2000, 7), Marx and Engels mobilized "metabolism" in an ontological manner; for them, human beings, like society, were an integral part of nature, albeit a particular and radically distinct part. The original German word for metabolism is *Stoffwechsel*, which literally translates as "change of matter." This simultaneously implies circulation, exchange, *and* the transformation of material elements. When matter moves, it becomes "enrolled" in associational networks that generate qualitative changes and qualitatively new assemblages. The newly produced "things" embody and reflect the processes of their making (through a process of internalization of dialectical relations), yet at the same time they differ radically from their constituent relational parts.

Every metabolized thing embodies the complex processes and heterogeneous relations of its past making. As it does so, it enters (or becomes enrolled) — in its turn and in its own unique way — into new assemblages of metabolic transformation. These dynamic heterogeneous assemblages form a circulatory process (though not necessarily closed). Under conditions of generalized commodity production, that process takes the form of circulation of commodities (and of non-commodified or partly commodified derivatives such as pollutants, garbage, CO_2, and the like) and the circulatory reverse flow of capital (as embodied dead labour in the form of past metabolic transformations). Encompassed by all of this are the particular power geometries associated with such socio-technical arrangements. According to Foster (2000), this processual metabolism is central to Marx's political economy; in a different way, it is also directly implicated in the circulation of commodities and, consequently, of money: "The economic circular flow then was closely bound up, in Marx's analysis, with the material exchange (ecological circular flow) associated with the metabolic interaction between human beings and nature" (ibid., 157–58).

Indeed, under capitalist social relations, the metabolic production of use values operates in and through specific control and ownership relations and in the context of the mobilization of both nature and labour to produce commodities (as forms of metabolized hybrid socio-natures) with an eye toward the realization of the embodied exchange value. The circulation of capital as value in motion is, then, the combined metabolic transformations of socio-natures in and through the reverse circulation of money as capital under social relations that combine the mobilization of capital, nature or dead labour, and labour power. In this metabolic process, new socio-natural forms, including the transformation of labour power as living labour, are constantly being produced as moments and things (see Grundman 1991; Benton 1989, 1996; Burkett 1999; Foster 2000). Whether we consider the production of dams, the re-engineering of rivers, the delivery of potable water, the management of biodiversity hot spots, the transfiguration of DNA codes, the cloning of species, the recycling of computer components by children in India, the trading of CO_2 emissions after Kyoto, the cultivation of potatoes (genetically modified or not), or the raising of a skyscraper, we find that all of these things testify to the particular assocational power relations through which socio-natural metabolisms are organized (in terms of property and ownership regimes, production or assemblaging activities, distributional arrangements, and consumption patterns).

Historical-geographical materialism offered a view of the world that unified the natural and the social while critiquing radically the "modern" separation of "society" from "nature" (Schmidt 1971; Smith 1984; Benton 1989). In fact, Bruno Latour's call in *We Have Never Been Modern* (1993) for us to reconnect the two poles that have been severed by modernity had already been sounded by Marx in *Grundrisse:* "It is not the unity of living and active humanity the natural, inorganic conditions of their metabolic exchange with nature, and hence their appropriation of nature, which requires explanation, or is the result of a historic process, but rather the separation between these inorganic conditions of human existence and this active existence, a separation which is completely posited only in the relation of wage labour and capital" (1973 [1858], 489).

Yet by focusing on the labour process as mere social process (as was and is the case for most of modern sociology, including Marxist sociology), some Marxist analysis—especially during the twentieth century—tended to replicate the very problem it meant to address. The connections referred to above were ignored rather than taken as the "space" for politics, for struggle, for prefiguring radical socio-ecological transformation and realizing alternative socio-natural relations. In other words, mainstream economics forgot the natural foundations of economic life (only to rediscover them recently under the guise of environmental economics); and at the same time, much of Marxist theory became an exclusive "social" theory rather than a socio-ecological one. Put simply, the overemphasis on social relations under capitalism that characterized much of Marxist (and other) social analysis tended to ignore or divert attention from the material and socio-physical metabolic relationships and their fantasmagoric representations and symbolic orderings. The result was a partial blindness in the twentieth-century social sciences to questions of political ecology and socio-ecological metabolisms.

Some recent approaches to the society–nature conundrum, such as actor network theory (see Latour 2004), have provided a new grammatical apparatus that has "profoundly revitalized empirical studies of human–nature–technology relations…But…it remains important that we incessantly raise the question…why are 'things as such' produced in the way they are—and to whose potential benefit" (Kirsch and Mitchell 2004, 20–21). While a historical-materialist mobilization of metabolism might begin to shed light on the production of socio-natural entities, this must be fused with another, equally central metaphor and material condition, one that is closely related to metabolism—namely, circulation.

The Invention of Circulation
The notion of "circulation" began to gain greater and wider currency in the natural and social sciences alongside the notion of "metabolism." When William in 1628 formulated his ideas of the double circulation of blood in the human body's vascular system, a revolutionary insight came into being that would permeate and dominate, both metaphorically and materially, engineering, academic practice, and everyday life for centuries to come. By the end of Harvey's century, medical practice had accepted the idea of the circulatory system; this led, among other things, to a profound redefinition of the body. In the nineteenth century the metabolic circulation of chemical substances and organic matter (see von Liebig's contribution above) became increasingly accepted; later on, this would form the basis of modern ecology. Ideas about the circulation and metabolism of matter became the two central metaphors for capturing processes of socio-natural change.

The term "circulation" to refer to the movement of money in a national economy established itself within a generation of Harvey's discovery (A.D. Harvey 1999). Thomas Hobbes, in *Leviathan* (1651), for example, had already compared the problems of a government that was unable to raise sufficient tax revenue to "an ague; wherein, the fleshy parts being congealed, or by venomous matter obstructed, the veins which by their natural course empty themselves into the heart, are not, as they ought to be, supplied from the artery, whereby there succeedeth at first a cold contraction, and trembling of the limbs; and afterwards a hot, and strong endeavour of the heart, to force the passage of the blood" (ibid.). Francis Bacon, in "Of Empire," wrote that merchants "are vena porta; and if they flourish not, a kingdom may have good limbs, but will have empty veins, and nourish little" (ibid.).

By the eighteenth century the concept of circulation had embedded itself in many sciences, in reference to everything from the flow of sap in plants to the circulation of matter in chemical reactions (Teich 1982). Circulation became a dominant metaphor after the French Revolution: ideas, newspapers, gossip, and—after 1880—traffic, air, and power all were said to circulate. After around 1750, wealth and money were perceived as circulating and were spoken of as though they were liquids, flowing incessantly to nourish a process of accumulation and growth. Society came to be imagined as a system of conduits (Sennett 1994). Montesquieu in *Lettres Persanes* (1973 [1721], 117) speaks of "the more 'circulation' the more wealth"; and in *l'esprit des lois* (1995 [1689–1755]) of "multiply[ing] wealth by increasing 'circulation.'" Rousseau (1766) refers to "this useful and fecund circulation that enlivens all society's labour" and to "a 'circulation of labour' as one speaks of the circulation of the money" (cited in Illich 1986). Intricate mechanical contraptions were constructed to mimic national economic dynamics as circuits of conduits, valves, and connections through which money and goods flowed incessantly. And, of course, by the mid-nineteenth century the flâneur—dandy, artist, detective, stroller, the favourite literary character of Baudelaire and later of Walter Benjamin—was well represented and theorized as an object of circulation within urban space. In the process, "circulation" became less closely identified with closed circular movement and more strongly with change, growth, and accumulation. In much the same way that von Liebig discovered the mechanisms of metabolism by considering the "metabolic rift," circulation acquired greater explanatory power precisely once it came to be viewed as integral to processes of transformation.

Adam Smith and, in particular, Karl Marx conceived a capitalist economy as a metabolic system of circulating money and commodities, both of which were carried by and structured through social interactions and relations. Accumulation depends on the swiftness with which money circulates through society. Each hiccup, stagnation, or interruption of circulation may unleash the infernal forces of devaluation, crisis, and chaos. Society's wealth and the power relations on which wealth is constructed are seen as bound up with and expressed by the "circulation speed" of money in all its forms (capital, labour, commodities). Later, David Harvey (1985) would analyze the circulation of capital and its urbanization as a *perpetuum mobile* channelled through a myriad of ever-changing production, communication, and consumption networks. The development of circulating money as the basis of material life, and the relations of domination and exclusion through which the circulation of money is organized and maintained, together have shaped this "urbanization of capital." Of course, von Liebig and Marx insisted that metabolic circulation was a process of destruction as well as of creation, one that harboured both enabling and disabling possibilities—a perspective that was largely lost in twentieth-century ecology and economics.

By the mid-nineteenth century some British architects were mobilizing the metaphor of circulation when speaking of inner cities. It was Sir Edwin Chadwick who first formulated the ideology of circulating waters, in 1842, when he presented a report on the sanitary conditions of the labouring population of Great Britain. In that report he imagined the new city as "a social body through which water must incessantly circulate, leaving it again as dirty sewage." Water, he surmised, ought to "circulate" through the city without interruption to cleanse it of sweats, excrements, and wastes (Vigarello 1988); unless water constantly circulates through the city, pumped in and channelled out, its interior space can only stagnate and rot. This representation of urban space as constructed in and through perpetually circulating flows of water is conspicuously similar to imagining the city as a vast reservoir of perpetually circulating money. Like the individual body and bourgeois society, the city, too, was now being described as a network of pipes and conduits. The brisker the flow, the greater the wealth, the health, and the hygiene of the city would be (Gandy 2004). In Chadwick's time new principles of city planning and policing were emerging based on the medical metaphors of "circulation" and "flow." The health of the body became the comparison against which the greatness of cities and states would be measured. The "veins" and "arteries" of the new urban design were to be freed from all sources of possible blockage (Sennett 1994, 262–65; Corbin 1994).

Urban Space as Spaces of Movement
With circulation as a metabolic process firmly established as practice and as a solid representation of socio-ecological change, attention quickly moved from metabolism and circulation to "speed"—in other words, to the "movement of movement." Metabolic circulation of the kind analyzed by Marx, and now firmly rooted in generalized commodity production, exchange, and consumption, was increasingly subjected to the socially

constituted dynamics of the capitalist market economy, in which the alpha and omega of the metabolic circulation of socio-ecological assemblages was the desire to circulate money as capital (Douglas 2004).

The creation of urban space as the space of the movement of people, commodities, and information radically altered the choreography of the city. Places and spaces became less and less shared; motion devalued or threatened to devalue place; connections were lost, identities reconfigured, and attachments broken down. Yet at the same time, the accumulation of movement and of capital signalled an intensified and accelerated accumulation of new urbanized natures, metabolized through metabolic vehicles that spun intricate networks and conduits. The urbanization of nature led to a spiralling accumulation of unstable socio-natural assemblages; at the same time, the components of these assemblages became radically disassociated from their geographical origins as speed, movement, and mobility—somewhat ironically—rendered the fields of vision and connections more opaque, transient, and partial. The city had turned into a metabolic vehicle, yet in the urban or modern imagination, the rift between nature and the social became deeper than ever.

Take, for example, the cybernetic networked service economies and cultures that are celebrated as the lifeblood of global cities. Many millions of computers are now linked together in a cyberspatial urban hyperreality; meanwhile, the accelerating turnover of the hardware to sustain this networked culture is generating spiralling volumes of e-waste. This is rapidly becoming the largest waste stream in the world, one that is globally organized and that produces massively uneven socio-spatial conditions, while radically reconstituting the socio-natural environment:

[E-waste] is not only of quantity but also one born of toxic ingredients—such as the lead, beryllium, mercury, cadmium, hexavalent chromium, and brominated-flame retardants that pose extraordinary occupational and environmental health threats...The estimated 315 million computers that became obsolete between 1997 and 2004 contain a total of more than 1.2 billion pounds of lead ...Computers and other electronics constitute a significant component of the physical and communicative infrastructure of cities. Given the level of natural resources required to produce these commodities, and the ecological damage that results from their production and disposal, e-waste is a symptom of the problematic of the human–nature interactions inherent in urban spaces. An estimated 80 percent of the US's computer waste collected for recycling is exported to Asia, where it is known to be dumped and recycled under very hazardous conditions. Environmental activists have called this "toxic colonialism" and a "global environmental injustice." (Pellow 2006, 231)

The global metabolic circulation of the cyberspace economy and the socio-natural power relations inscribed in this remain hidden. While the focus is on speed and high-tech networks, the material socio-environmental connections and the uneven power relations that produce them remain blatantly invisible. The surface of the wired city hides a disturbing socio-ecological underbelly.

(Hybrid) Natures and (Cyborg) Cities

When we mobilize the twin vehicles of metabolism and circulation from a historical-materialist epistemological perspective, the binary construction of "nature" and "society" that has characterized much of the modern scientific and cultural tradition abruptly disappears. Metabolic circulation, then, is the socially mediated process of environmental—including technological—transformation and transconfiguration, through which all manner of "agents" are mobilized, attached, collectivized, and networked. The heterogeneous assemblages that emerge, as moments in the accelerating and intensifying circuitry of metabolic vehicles, are central to a historical-geographical materialist ontology and imagination. A dialectical approach recognizes the radical heterogeneity of humans and non-humans enrolled in socio-metabolic processes within an assemblage while also recognizing the social, cultural, and political power relations embodied relationally in these socio-natural or technonatural imbroglios. The production of (entangled) things through metabolic circulation is necessarily a process of fusion, of the making of "heterogeneous assemblages," of constructing longer or shorter networks.

These assemblages of humans and non-humans, of dead labour and inert materials, are nevertheless reminiscent of the "hybrids" and "cyborgs" of, respectively, Bruno Latour and Donna Haraway.

However, some qualifications need to be noted when applying these concepts. First, "hybridity" and "cyborg" can be misleading terms: the bracketing of both in this section's title points to the "excess of meaning" that results when the city is coded as one or the other. While intuitively attractive, both terms suggest—which they should not—a process of "dirty" mixing, an ambiguous fusion of things that can be ontologically separated and "purified." In fact, natures and cities are always already heterogeneously constituted, the product of actants in metabolic circulatory processes.

Second, Haraway asks penetrating questions about why cyborgs are produced the way they are and about the power relations inscribed in these imbroglios; Latour does not. Latour calls for a new socio-natural constitution, for a "reassembling of the social" in a way that recognizes the social acting of non-human things, and he does so in ways apparently similar to Haraway; yet he does not just ignore questions of power, but rejects them sharply as a fruitful approach to excavating socio-natural imbrogilios (such as e-waste, potable water, and nuclear energy). For him, the key issue is how to transform the "constitutional" arrangements through which human and non-human actants become mobilized or enrolled (Latour 2004).

In sum, Latour defends a democratic republic of heterogeneous associations, whereas Haraway's perspective emerges from a radically different ontological position. A deep ontological divide is evident here. As Benedikte Zitouni (2004) convincingly argues:

> Haraway views any entity as an *embodiment* of relations, an *implosion*, the threads of which should be teased apart in order to understand it. Whereas Latour views any entity as *a piece of matter* that is continuously affected and that contracts links with a larger networks *that allows it to live, to be.* On the one hand, the entity *crystallizes* the network; on the other hand the entity is *supported* by the network. Haraway studies the network in order to define the entity; Latour studies that same network in order to define the entity's consistency and persistence ... Dialectics, congealment, crystals, prisms, representations are not possible tools any longer for urban studies but instead we view pieces of matter, of any kind, that act, react and interact with one another, that gain their consistency, persistence and existence or lose them through the affects and links to other agents. Power differences and inequality can no longer be stated as such, as a departure point into the city but have to be explained through the many actions and relations between objects, humans and non humans. There is nothing *behind* any space or agent, only attachments *aside* of it that make it stronger or weaker, allow it to exist or lead it to perish. (2004, 8)

It is in this latter sense that we wish to see the city as a metabolic circulatory process that materializes as an implosion of socio-natural and socio-technical relations organized through socially articulated networks and conduits whose origin, movement, and position are articulated through complex political, social, economic, and cultural relations. These relations are invariably infused with myriad configurations of power that saturate material practices, symbolic ordering, and imaginary (or imagined) visions.

Yet little attention has been paid so far to the urban as a flow or a process of socio-ecological *change*. In other words, the view that a city is a process of environmental production, sustained by particular sets of socio-metabolic interactions that shape the urban in distinct, historically contingent ways — a socio-environmental process that is deeply caught up with socio-metabolic processes operating elsewhere and producing profoundly uneven socio-ecological conditions — rarely grabs the headlines. It is to this mode of thinking about the city that we turn next.

The Urban as a Flow of Socio-Ecological Change

The political-ecological history of any city can be written from the perspective of the need to urbanize and domesticate nature and the concomitant need to push the ecological frontier outward as the city expands (Swyngedouw 2004). As such, urban political-ecological processes produce both a new urban socio-nature and a new rural one. The city's growth and nature's urbanization are both closely associated with successive waves of ecological transformation, with the socio-ecological organization of metabolic processes as defined above, and with the extension of urban socio-ecological frontiers. Local, regional, and national socio-natures are combined with engineering narratives, economic discourses and practices, land speculation, geopolitical tensions, and global money flows. This metabolic circulation is deeply entrenched in the political ecology of the local and national state, the international divisions of labour and power, and local, regional, and global socio-natural networks and processes.

However, while socio-environmental (both social and physical) qualities may be enhanced in some places and for some people, they often lead to a deterioration of social and physical conditions and qualities elsewhere (Peet and Watts 1993; Keil and Graham 1998; Laituri and Kirby 1994), both within cities and between cities and other often very distant places. A focus on the uneven geographical processes inherent in the production of urban environments allows a better understanding of socio-ecological urbanization. Perpetual change and an ever-shifting mosaic of environmentally and socio-culturally distinct urban ecologies—ranging from the manufactured and manicured landscaped gardens of gated communities and high-tech campuses to the ecological war zones of depressed neighbourhoods, with their lead-painted walls and asbestos-covered ceilings, their waste dumps and pollutant-infested vacant lots—still shape the choreography of capitalist urbanization.

The environment of the city is deeply caught up in this dialectical process; and environmental ideologies, practices, and projects are part and parcel of this urbanization of nature (Davis 2002; Keil 2003). From this perspective, there is no such thing as an unsustainable city in general; rather, there are a series of urban and environmental processes that harm some social groups while benefiting others (see Swyngedouw and Kaika 2000). It follows that a just urban socio-environmental perspective must always consider who gains and who pays and ask serious questions about the multiple power relations—and the networked and scalar geometries of those relations—whereby deeply unjust metabolic processes are produced and maintained. In other words, environmental transformations are not independent of class, gender, ethnicity, and other power struggles. These metabolisms generate socio-environmental processes that are both enabling (for powerful individuals and groups) and disabling (for marginalized individuals and groups). They precisely produce positions of empowerment and disempowerment. Because these relations form under and can be traced directly back to the crisis tendencies inherent to neoliberal forms of capitalist development, the struggle against exploitative socio-economic relations necessarily fuses with struggles to bring about more just urban environments (Bond 2002; Swyngedouw 2005).

Processes of socio-environmental change and the reconfiguration of metabolic circulatory arrangements are, therefore, never socially or ecologically neutral. The result is conditions wherein particular trajectories of socio-environmental change undermine the stability of humans and non-humans (and the local ecologies of which they are part) in some places, while sometimes enhancing their "sustainability" elsewhere. In sum, a political-ecological examination of the urbanization process reveals the inherently contradictory nature of metabolic change, and the technonatural "metabolic vehicles" of that change tease out the inevitable conflicts (or the displacements thereof) that infuse socio-environmental change (see Swyngedouw, Kaika, and Castro 2002). It is this nexus of power and the social actors deploying or mobilizing these power relations that ultimately decide who will have access to or control over, and who will be excluded from access to or control over, resources or other components of the environment.

These power relations shape the socio-natural configurations of the urban environments in which we live. Because the power-laden socio-ecological relations that go into the formation of urban environments constantly shift among groups of human and non-human actors as well as among spatial scales, historical-geographical insights into these ever-changing urban configurations are necessary for the sake of considering the future evolution of urban environments.

Developing a Political Ecology of the Urban

An urban political-ecological perspective permits new insights in the urban problematic and opens new avenues for recentring the urban as the pivotal terrain for ecopolitical action. To the extent that emancipatory urban politics reside in acquiring the power to produce urban environments in line with the needs and aspirations of those inhabiting these spaces, and in the capacity to produce the physical and social environment in which one dwells, the question of whose nature is or becomes urbanized must be at the forefront of any radical political action. This suggests a research agenda; it also opens up a political platform that may indicate how to democratize the politics through which cities are produced as both enabling and disempowering sites of living for humans and non-humans. "Urbanizing" the environment, then, is a project of social and physical environmental construction that actively produces the urban (and other) environments we wish to inhabit today.

A number of recent publications have begun to address this problematic. William Cronon (1991) in *Nature's Metropolis* tells the story of Chicago from the vantage point of the socio-natural processes that transformed both city and countryside, that generated the specific political ecology which shaped the transformation of the Midwest and produced a particular American socio-nature. While eerily silent about the myriad struggles that have infused this process (African-American, women's, and workers' organizations and struggles are notoriously absent from or marginalized in his narrative), the book points to interesting and powerful approaches to a political ecology of the urban.

Mike Davis (1990), in *City of Quartz* and other recent publications

(Davis 1996, 1998), in and through the dialectics of Los Angeles's urbanization process, suggests how nature and society become materially and discursively constructed and how multiple social struggles infuse and shape this process in deeply uneven, exclusive, and empowering/disempowering ways. For him, homelessness and racism, combined with pollution, earthquakes, and water scarcity, are the most acute socio-ecological problems that have been produced through the particular form of postindustrial capitalist development that has shaped Los Angeles as the Third World Megalopolis. Davis's history of LA's urbanization describes how deserts lands were socio-ecologically transformed; how an orchard socio-nature was manufactured; and how, later on, "silicon" landscapes were constructed; how at the same time ever larger and more distant watersheds were captured, controlled, and urbanized; how speculators pushed ever outward the frontier of "developable" land; and how an immensely contested and socially significant web (in terms of access and exclusion; empowerment/disempowerment) of national laws, regulations, and engineering projects came to be choreographed (Worster 1985; Gottlieb and Fitzsimmons 1991). Of course, as the deserts bloomed, ecological and social disaster struck: water scarcity, pollution, congestion, and lack of sewage disposal combined with mounting economic and racial tensions and a rising environmentalism (O'Connor 1998, 118; Keil and Desfor 1996; Keil 1998). The rhetoric of disaster, risk, and scarcity often provided the discursive vehicles by means of which power brokers could constantly reinvent their boosterist dreams. Pictures of a simulacrum of drought, scarcity, and redesertification generated a spectacularized vision of the dystopian city, whose fate — people now believe — is directly related to those administrators, engineers, and technicians who make sure that the taps keep flowing and that land keeps being "developed." The hidden stories of pending socio-ecological disaster serve as a crucible in which local, regional, and national socio-natures are combined with engineering narratives, land speculation, and global flows of water, wine, and money.

Matthew Gandy (2002) excavates with great skill and in exquisite detail the reworking of nature in New York City, a reworking that is simultaneously material and physical as well as embedded in political, social, and cultural framings of nature. At the same time, the myriad power relations and political strategies that infuse the socio-environmental metabolism of New York's socio-nature are meticulously excavated and brought onto centre stage in the reconstruction of contemporary New York as a cyborg city.

In my own recent work (see Swyngedouw 2004), I have charted how the flow of water in its material, symbolic, political, and discursive constructions embodies and expresses precisely how the "production of nature" is both arena and outcome of the tumultuous reordering of socio-nature in ever-changing and intricate ways. I treat this flow of water as a socio-environmental metabolic process and its historical-geographical production as an entry point to excavate the processes of modern urbanization in Guayaquil, Ecuador. Guayaquil, Ecuador's largest and most powerful city, on the Pacific coast, suffers from a seriously socially uneven access to potable urban water, like many other cities in developing countries. Of its 2 million inhabitants, 38 percent do not have access to piped potable water and depend on private vendors, who sell water at massively inflated prices. Publicly supplied water costs about three cents for 1000 litres; private water vendors charge three dollars. As a result, an intense social and political struggle, enacted on bodily, neighbourhood, urban, regional, national, and international scales, has been unfolding over access to and control over the city's water resources. The uneven power relations that have shaped urbanization in Guayaquil are in this way etched into the circulation of urban H_2O. The city of Guayaquil and the urbanization of potable water developed on the basis of successive ecological conquests and the appropriation of rents, from agricultural produce or the pumping of oil, through which money was constantly recycled and nature became urbanized.

Urbanization inserted water squarely into the circulation process of money and its associated power relations and class differentials. With each round of accumulation, based successively on cocoa, bananas, and oil, the territorial scale of the socio-ecological complex changed and the scalar geographies of political power became rearticulated. A new configuration of elites would each time reorganize the socio-ecological configuration of the urbanization process and shape the hydrosocial networks according to its own interests and logic. In the process, the hydrosocial flow became transformed and restructured, culminating in heavily lopsided access to water.

The production of the city as a cyborg, excavated through the analysis of the circulation of hybridized water, opens up a new arena for thinking and acting in the city—an arena which is neither local nor global but which weaves a network that is always deeply localized even while extending its reach over a certain scale, a certain spatial surface. The tensions, conflicts and forces that flow with water through the body, the city, the region, and the globe show the cracks in the lines, the meshes in the net, the spaces and plateaus of resistance and of power.

Conclusion: The Urbanization of Nature

"Metabolism" and "circulation" permit us to excavate the socio-environmental basis of the city's existence and its change over time. The socio-naturally "networked" city can be understood as an intricate socio-environmental process, one that perpetually transforms the socio-physical metabolism of nature. Nature and society in this way combine to form an urban political ecology, a hybrid, an urban cyborg that combines the powers of nature with those of class, gender, and ethnic relations. In the process a socio-spatial fabric is generated that privileges some and excludes many, thereby generating significant socio-environmental injustices. Nature, then, is integral to the political ecology of the city and needs to be addressed in these terms. The urbanization of nature, though generally portrayed as a technological/engineering problem, is in fact as much part of the politics of life as any other social process. It is essential to recognize this political meaning of nature if sustainability is to be combined with just and empowering urban development — urban development that returns the city and the city's environment to its citizens. In other words, socio-ecological metabolisms are inherently political and thus integral to any political or social project. Political visions are, therefore, necessarily also ecological visions; any political project must also be an environmental project, and vice versa.

Environmental and social changes co-determine each other. Processes of socio-environmental metabolic circulation transform both social and physical environments and produce social and physical milieux (such as cities) with new and distinct qualities. In other words, environments are combined socio-physical constructions that are actively and historically produced in terms of both social content and physical features. Whether it is urban parks, urban natural reserves, or skyscrapers, all express fused socio-physical and technonatural processes that contain and embody particular metabolic and social relations.

There is, consequently, nothing unnatural about produced environments such as cities, genetically modified organisms, dammed rivers, or irrigated fields. Produced environments are specific historical results of socio-environmental processes. The urban world is a cyborg world, part natural and part social, part technical and part cultural, and with no clear boundaries, centres, or margins. All socio-spatial processes are invariably also predicated on the circulation and metabolism of physical, chemical, and/or biological components. Non-human "actants" play an active role in mobilizing socio-natural circulatory and metabolic processes. It is these circulatory conduits that link often distant places and ecosystems together and that permit us to relate local processes with broader socio-metabolic flows, networks, configurations, and dynamics.

These socio-environmental metabolisms produce a series of both enabling and disabling social and environmental conditions. Environmental (both social and physical) qualities may be enhanced in some places and for some humans and non-humans, but they also often lead to deteriorating social, physical, and/or ecological conditions elsewhere. Processes of metabolic change are never socially or ecologically neutral. This results in conditions under which particular trajectories of socio-environmental change undermine the stability or coherence of some social groups, places, or ecologies, while their sustainability elsewhere may be enhanced.

In sum, the political-ecological examination of the urbanization process reveals the inherently contradictory nature of metabolic circulatory change and teases out the inevitable conflicts (or the displacements thereof) that infuse socio-environmental change. Social power relations (be they material or discursive, economic, political, and/or cultural) through which metabolic circulatory processes take place are especially important. It is these power relations through which human and non-human actors become enrolled, and the socio-natural networks that carry them ultimately decide who will have access to or control over, and who will be excluded from access to or control over, resources and other components of the environment, and who or what will be positively or negatively enrolled in such metabolic imbroglios. These power relations, in turn, shape the particular social and political configurations and the environments in which we live.

Questions of socio-environmental sustainability are therefore fundamentally political questions. The politics of socio-ecological transformations tease out who (or what) gains from and who pays for, who (or what) benefits from and who suffers (and in what ways) from, particular processes of metabolic circulatory change. It also seeks answers to questions about what or who needs to be sustained and how this can be achieved. This requires us to unravel the nature of the social relationships that unfold between individuals and social groups and how these, in turn, are mediated by and structured through processes of ecological change. In other words, urban political ecology demonstrates how socio-ecological "sustainability" can only be achieved through a democratically controlled and organized process of socio-environmental (re)construction.

References

Althusser, L. 1969. *For Marx.* London: Verso.

Baeten, G. 2000. "Tragedy of the Highway: Empowerment, Disempowerment, and the Politics of Sustainability Discourses and Practices." *European Planning Studies* 8, no. 2: 69–86.

Benton, T., ed. 1996. *The Greening of Marxism.* New York: Guilford.

———. 1989. "Marxism and Natural Limits: An Ecological Critique and Reconstruction." *New Left Review* 178: 51–86.

Bond, P. 2002. *Unsustainable South Africa.* London: Merlin.

Burkett, P. 1999. *Marx and Nature—A Red and Green Perspective.* New York: St. Martin's.

Chadwick, E. 1887. *The Health of Nations,* 2 vols. London: Richardson.

———. 1842. *Report on the Sanitary Conditions of the Labouring Population of Great Britain.* London: B.P.P., Vol. 26.

Corbin, A. 1994. *The Foul and the Fragrant.* London: Picador.

Cronon, W. 1991. *Nature's Metropolis—Chicago and the Great West.* New York: Norton.

Davis, M. 2002. *Dead Cities.* New York: New Press.

———. 1998. *Ecology of Fear: Los Angeles and the Imagination of Disaster.* New York: Metropolitan.

———. 1996. "How Eden Lost Its Garden: A Political History of the Los Angeles Landscape." In *The City—Los Angeles and Urban Theory at the End of the Twentieth Century,* ed. A.J. Scott and E.W. Soja. Berkeley: University of California Press. 160–85.

———. 1990. *City of Quartz: Excavating the Future of Los Angeles.* London: Verso.

Douglas, I.R. 2004. "The Calm before the Storm: Virilio's Debt to Foucault, and Some Notes on Contemporary Global Capital." http://proxy.arts.uci.edu/~nideffer/_SPEED_/1.4/articles/douglas.html.

Fisher-Kowalski, M. 2003. "On the History of Industrial Metabolism." In *Perspectives on Industrial Ecology,* ed. D. Bourg and S. Erkman. Sheffield: Greenleaf. 33–45.

———. 1998. "Society's Metabolism: The Intellectual History of Material Flow Analysis, Part I, 1860–1970." *Journal of Industrial Ecology* 2, no. 1: 61–78.

Foster, J.B. 2000. *Marx's Ecology: Materialism and Nature.* New York: Monthly Review.

Gandy, M. 2005. "Cyborg Urbanization: Complexity and Monstrosity in the Contemporary City." *International Journal of Urban and Regional Research* 29, no. 1: 26–49.

———. 2004. "Rethinking Urban Metabolism: Water, Space, and the Modern City." *City: Analysis of Urban Trends, Culture, Theory, Policy, Action* 8, no. 3: 371–87.

———. 2002. *Concrete and Clay—Reworking Nature in New York City.* Cambridge, MA: MIT Press.

Godelier, M. 1986. *The Mental and the Material.* London: Verso.

Gottlieb, R., and M. Fitzsimmons. 1991. *Thirst for Growth.* Tucson: University of Arizona Press.

Grundman, R. 1991. *Marxism and Ecology.* Oxford: Clarendon.

Haraway, D. 1991. *Simians, Cyborgs, and Women—The Reinvention of Nature.* London: Free Association.

Harvey, A.D. 1999. "The Body Politic: Anatomy of a Metaphor." *Contemporary Review,* August. http://articles.findarticles.com/p/articles/mi_m2242/is_1603_275/ai_55683940.

Harvey, D. 1985. *The Urbanization of Capital.* Oxford: Blackwell.

Harvey, W. 1628. Exercitatio *Anatomica de Motu Cordis et Sanguinis in Animalibus.* Francofurti: Sumptibus Gulielmi Fitzeri.

Illich, I. 1986. *H₂O and the Waters of Forgetfulness.* London: Marion Boyars.

Jameson, F. 2002. *A Singular Modernity.* London: Verso.

Kaika, M., and E. Swyngedouw. 1999. "Fetishising the Modern City: The Phantasmagoria of Urban Technological Networks." *International Journal of Urban and Regional Research* 24, no. 1: 120–38.

Keil, R. 2003. "Urban Political Ecology." *Urban Geography* 24, no. 8: 723–38.

———. 1998. *Los Angeles: Globalization, Urbanization, and Social Struggles.* World Cities Series. New York: Wiley.

Keil, R., and G. Desfor. 1996. "Making Local Environmental Policy in Los Angeles." *Cities* 13, no. 5: 303–13.

Keil, R., and J. Graham. 1998. "Reasserting Nature: Constructing Environments after Fordism." In *Remaking Reality: Nature at the Millennium,* ed. B. Braun and N. Castree. London: Routledge. 100–25.

Kirsch, S., and D. Mitchell. 2004. "The Nature of Things: Dead Labor, Nonhuman Actors, and the Persistence of Marxism." *Antipode* 36, no. 4: 687–706.

Laituri, M., and A. Kirby. 1994. "Finding Fairness in America's Cities? The Search for Environmental Equity in Everyday Life." *Journal of Social Issues* 50, no. 3: 121–39.

Latour, B. 2004. *Politics of Nature: How to Bring the Sciences into Democracy.* Cambridge, MA: Harvard University Press.

———. 1993. *We Have Never Been Modern.* London: Harvester Wheatsheaf.

Levins, R., and R. Lewontin. 1985. *The Dialectical Biologist.* Cambridge, MA: Harvard University Press.

Liebig von, J. 1842. *Animal Chemistry: or, Organic Chemistry in Its Application to Physiology and Pathology.* Edited from author's manuscript by William Gregory. With additions, notes, and corrections by Dr. Gregory and John W. Webster. Facsimile of the Cambridge edition of 1842. New York: Johnson Reprint.

———. 1840. *Principles of Agricultural Chemistry, with Special Reference to the Late Researches Made in England.* London: Walton and Maberly.

Luke, T.W. 2003. "Global Cities vs. 'Global Cities.'" *Studies in Political Economy* 7, no. 1: 11–22.

———. 1996. "Liberal Society and Cyborg Subjectivity: The Politics of Environments, Bodies, and Nature." *Alternatives* 21: 1–30.

Marx, K. 1973 [1858]. *Grundrisse.* New York: Vintage.

———. 1970. *Capital,* Vol. 1. New York: Penguin.

Moleschott, J. 1857. *Der Kreislauf des Lebens.* Mainz: Von Zabern.

Montesquieu, C. de Secondat, baron de. 1995 [1689–1755]. *L'esprit des lois.* Paris: Nathan.

———. 1973 [1689–1755]. *Lettres Persanes. Edition établie et présentée par Jean Starobinski.* Collection Folio 475. Paris: Gallimard.

Norgaard, R. 1994. *Development Betrayed: The End of Progress and a Co-evolutionary Revisioning of the Future.* New York and London: Routledge.

O'Connor, J., ed. 1998. *Natural Causes: Essays in Ecological Marxism.* New York: Guilford.

Padovan, D. 2000. "The Concept of Social Metabolism in Classical Sociology." *Revista Theomai* 2. http://redalyc.uaemex.mx/redalyc/pdf/124/ 12400203.pdf.

Page, B. 2004. "Cyborg Apartheid and the Metabolic Pathways of Water in Lagos." Paper presented at the research colloquium "Re-Naturing Urbanization," held at Oxford University, June 30.

Peet, R., and M. Watts, eds. 1996. *Liberation Ecologies.* London: Routledge.

Pellow, D. 2006. "Transnational Alliances and Global Politics: New Geographies of Urban Environmental Justice Struggles." In *In the Nature of Cities: Urban Political Ecology and the Politics of Urban Metabolism,* ed. N. Heynen, M. Kaika, and E. Swyngedouw. London: Routledge. 226–44.

Schmidt, A. 1971. *The Concept of Nature in Marx.* London: New Left.

Sennett, R. 1994. *Flesh and Stone.* London: Faber and Faber.

Smith, N. 1984. *Uneven Development: Nature, Capital, and the Production of Space*. Oxford: Blackwell.

Swyngedouw, E. 2005. "Dispossessing H_2O — The Contested Terrain of Water Privatisation." *Capitalism, Nature, Socialism* 16, no. 1: 1–18.

———. 2004. *Social Power and the Urbanization of Water: Flows of Power*. Oxford: Oxford University Press.

———. 1996. "The City as a Hybrid — on Nature, Society, and Cyborg Urbanisation." *Capitalism, Nature, Socialism* 7, no. 1: 65–80.

Swyngedouw, E., and M. Kaika. 2000. "The Environment of the City or... the Urbanization of Nature." In *Companion to Urban Studies*, ed. G. Bridge and S. Watson. Oxford: Blackwell. 567–80.

Swyngedouw, E., M. Kaika, and E. Castro. 2002. "Urban Water: A Political-Ecology Perspective." *Built Environment* 28, no. 2: 124–37.

Teich, M. 1982. "Circulation, Transformation, Conservation of Matter, and the Balancing of the Biological World in the Eighteenth Century." *Ambix* 29: 17–28.

Vigarello, G. 1988. *Concepts of Cleanliness — Changing Attitudes in France since the Middle Ages*. Cambridge: Cambridge University Press.

Virilio, P. 1986. *Speed and Politics: An Essay on Dromology*. Cambridge, MA: Semiotext(e).

Worster, D. 1985. *Rivers of Empire: Water, Aridity, and the Growth of the American West*. New York: Pantheon.

Zitouni, B. 2004. "Donna Haraway and Bruno Latour: An Ontological Divide." Paper presented to conference "Technonatures II," School of Geography and the Environment, Oxford University, June 24.

Chapter Three

The Cellphone-in-the-Countryside: On Some of the Ironic Spatialities of Technonatures

Mike Michael

A central theme of this book is how the purifying rhetoric of much environmentalist discourse increasingly fails to capture the complex entanglements and constructions of technologies, natures, and social life. In this chapter I develop this theme by exploring some of the complexities that surround the cellphone[1] when considered in the seemingly alien context of the countryside. I begin with the assumption that neither "cellphone" nor "countryside" is a priori distinct: rather, my focus is on "the cellphone-in-the-countryside" — the hyphenation indicating that cellphone and countryside are partly co-constituted through their relations to each other (as well as through other relations). This co-constitution is both material and semiotic and entails a straddling and muddling of numerous dichotomies that have characterized Western modernist modes of thought (e.g., Latour 1993). As we shall see, these dichotomies include technology and nature, rural and urban, public and private, domesticated and wild, local and global, the qualitative and the quantitative, and safe and risky.[2]

I trace these entanglements with the aid of the concept of "ironic technonatural spatialities." "Technonatural," obviously enough, connotes the complex entanglements and constructions of the natural and technological, not least as these are respectively evoked in the figures of the "countryside" and the "cellphone." "Spatialities" refers to the way that "the cellphone-in-the-countryside" features in the literal making of spaces.

That is to say, I will try to demonstrate in this chapter how "the cellphone-in-the-countryside" serves at once to mediate and subvert such spaces as the private and the public, the rural and the city, the safe and the risky.[3]

I call these spatialities "ironic" because, in keeping with other recent treatments of spatiality (e.g., Whatmore 2002; J. Law 2004), I want to get at the way in which "behind" any one particular reading of a state of affairs can always be found another contrasting one (on irony, see for example Muecke 1969). Of course, "irony" typically refers to meaning, but in this instance I take meaning to be a matter of ordering that is not simply linguistic—that also heterogeneously entails corporeal and material associations.[4] As such, in the present instance, to talk of "ironic spatiality" is also to entertain material and corporeal irony wherein multiple and contrasting heterogeneous orderings (and disorderings) are co-present. For example, on one level it is claimed that "the cellphone-in-the-countryside" disrupts the countryside by importing urban values; on another level, tourists from the city are a source of revenue that makes such a countryside "sustainable."

Now, to address these concerns, we first need to consider how the cellphone has thus far been treated, not least in studies of everyday life and technology. As we shall see, and in keeping with the broad tenor of much of the sociology of everyday life, the cellphone is conceptualized largely in relation to urban settings, and primarily in terms of the ways in which it mediates social relations of one sort or another. By comparison, by looking at a few of the ironies that attach to the cellphone-in-the-countryside, we can begin to trace not only how the various dichotomies mentioned above are enacted, but also how these collapse to enable a rather different account of the technonatural complexities of the "countryside," where the cellphone also mediates relations between humans and the natural, that is to say, divergent instantiations of the technonatural.

In what follows I briefly discuss previous accounts of the cellphone read through some abiding concerns in the sociology of everyday life. I go on to consider the "cellphone-in-the-countryside," mainly through two examples that indicate how the capacities of the cellphone are problematized—or better still, ironized. I then expand this analysis to consider some broader socio-technical assemblages that enable cellphones to function in particular ways, not least ways that are associated with a range of technonatural risks. Finally, I reflect briefly on the politics of ironic technonatural spatialities.

Cellphones and Everyday Life

What would Henri Lefebvre and Michel de Certeau make of the cellphone? From these two key figures in the contemporary reinvigoration of the sociology of everyday life (see Gardiner 2000; Highmore 2002), two rather different accounts would be expected. From Lefebvre we might expect a critique of the cellphone's role in what he famously called "the bureaucratic society of controlled consumption" (Lefebvre 1968, 60). From this perspective, the virtual consumption entailed in the use of the cellphone detracts from the process of engaging with people who are physically co-present. In other words, people are so absorbed in their cellphones that "reciprocal forms of recognition" (Bull 2000) are eroded: people stop interacting with one another locally and thus the possibility of making politically potent social bonds is diminished. Yet at the same time, the cellphone mediates forms of mutual recognition with others-at-a-distance. This can serve as the basis of virtual-ish community—a community with the potential to generate a variety of political possibilities, including utopian ones (for a utopian view of this sort of potential for mobile ICTs, see Poster 2002; cf. Shields 1999). De Certeau (1984) would view the cellphone as a strategic mechanism for continued (panoptical) surveillance (certainly, that is how children and employees sometimes perceive it; see Haddon 2004). But the cellphone can also tactically subvert surveillance—that is, it can serve as the socio-technical grounds for a momentary ruse where one misplaces the phone or "forgets" to switch it on. To be sure, these concocted accounts leave something to be desired—not least in the way the cellphone is shoehorned into a pre-existing set of high-theoretical concerns at the expense of an engagement with the empirically complex uses of the cellphone (see Bennett 2004; Michael 2006).

In his preliminary outline of a sociology of the cellphone, Jim McGuigan (2004) sketches four generally appropriate approaches: social demography; political economy; conversation, discourse, and text analysis; and ethnography. He focuses on the last of these, not least because this can better represent "the nuances and complexities of everyday life" (2004, 51). In reviewing three studies, he notes how such ethnographies can be "interested"—portraying, albeit with subtlety, some of the ways that cellphones can reshape social relations and social interactions. Rereading McGuigan's analysis (and various other studies; e.g., Cooper et al. 2002; Katz and Sugiyama 2005; Persson 2001) through the idea of ironic spatiality, we can see that cellphones are caught up in the performance of social space—social space that is shot through with divergent overlying and underlying meanings (in the sense of heterogeneous orderings and disorderings outlined above). Thus the way people bodily position themselves when talking into their phones enacts the complexities of the boundary (if that is the correct term) between public and private spaces—there is a folding in of the public into the private, and vice versa, as people's privatizing movements are shaped by the physical presence of the public, even as those movements are used to signal (i.e., perform) privacy to that public. People struggle with the ironies of the cellphone as a fashion accessory and functional technical artifact—like walking boots (Michael 2000), cellphones can signify, in contrary ways, a fine sense of style and a fine sense of utility. Both these forms of cellphone consumption reflect an aspiration to a particular status (or a display of particular taste), whether as a consumer of fashionable goods or as someone whose persona is bound up with the consumption of goods that signal some form of studied non-consumption. Finally, we can note that the display of phones can serve as a marker of contactability and ready access to assistance in situations where people (notably women) feel at risk of physical attack; but at the same time, such displays leave people (not least younger people) at risk of theft and physical attack. Here is an irony: the cellphone is a dangerously desirable object of protection—it is, in other words, a risk-repulsing attractor of risk.

We have touched on some of the ironic spatialities out of which the cellphone emerges, and which it shapes in its trajectories through social space. But note that the emphasis here is on the "social." The cellphone—no doubt partly reflecting the legacy of the sociology of everyday life—is here portrayed as a "medium" through which social relations of various complex sorts are enacted. There are a number of points to unravel here. First, "social" refers more or less exclusively to urban settings. As with much sociology of everyday life, the city is seen as the primary empirical site in which the processes of everyday life are to be investigated (and conversely, it is through the sociology of everyday life that the city has come to be studied; see Amin and Thrift 2002).[5] As such, the urban seems to take precedence over the rural. Second, insofar as the cellphone serves in the production of social relations, this deflects from its role in the making of heterogeneous relations. Of course, we can nowadays take for granted that social relations are necessarily heterogeneous, but what interests us here is the precise ways in which cellphones mediate such heterogeneity. For example, to talk on a cellphone on a train is also to deal with varying volumes of train rattle, or the loss of signals as the train passes through tunnels. Rather, the cellphone, as a focus of analytic attention, should be seen as a heterogeneous artifact (or rather a socio-technical assemblage — see below) that helps pattern heterogeneous orderings and disorderings (see Michael 2000). Putting the first and second points together brings us to our third: the cellphone's heterogeneity is played out in the context of the "natural" and the "environmental." Cellphones also mediate technonatures insofar as they are instrumental in the everyday or mundane shaping of the boundary between the rural and the urban — or, rather, in the peculiar "unbordering" of urban and rural that comprises particular ironic spatialities of technonature.

Next we explore some ironic spatialities of technonature that might be associated with the cellphone.

Some Ironic Spatialities of Technonature

Let me begin with two differing accounts of the cellphone in the "countryside."

On January 25, 2005, *The Times* (London) ran a report on how walkers had used their cellphones to call for help from the mountain-rescue service in the Lake District, in the northwest of England. The article ran as follows:

Irresponsible mobile (cell) phone use is wasting mountain rescuers' time, claims team leader Stuart Hulse. And he should know.

The distress call from the mobile phone of a couple stranded up a mountain in the Lake District sounded serious. "We are lost in the mist," said an anxious voice. "My wife is very frightened. Please come and find us." As the message continued, the mountain rescue team listened incredulously. "And could you send a helicopter?" asked the caller. "We have a dinner date at 7pm which we really don't want to miss."

Stuart Hulse, who has been a mountain rescuer for nearly 40 years, can tell many more stories like this. Such as the recent case of a group of 11 professionals on a management bonding weekend who rang claiming to be lost in a valley.

The mountain rescuers were scrambled and located them within 20 minutes. It transpired that, although the group had gone slightly out of their way, they had no transport and wanted to be driven back to their accommodation. "It would have cost them a fortune to get three taxis to take them all back to their base," says Hulse. "So we ferried them back over the passes." Mountain rescue today, he adds angrily, is increasingly becoming a "nannying and taxi service."

Like many mountain rescuers, Hulse is in no doubt where the root of this problem lies. Mobile phones. Already considered the scourge of city dwellers, the ubiquitous mobile is now creating havoc in the countryside. It has, he says, created a new breed of climber and fellwalker who no longer sees the need to go out equipped with a torch, a compass, a map or even adequate clothing because he knows that help is just a phone call away. Common sense and personal responsibility have been replaced by complacency and the expectation of instant gratification. (http://www.timesonline.co.uk/article/ 0,7-1454604_1,00.html. Accessed July 19, 2006)

The Wales Tourist Board has, since 2004, been running an advertising campaign promoting Wales as "The Big Country." Wales is represented as a bastion of "nature" and "heritage" in the face of the supposedly less desirable onslaughts of modernity and globalization. Thus, in one poster, the reader is presented with a score line of the number of castles in Wales (461) against the number of Starbucks (6). In another—the one that interests us here—over a spectacular view of Snowdonia, there are printed the words "Area of Outstandingly Bad Mobile (Cellphone) Reception"— a parody of the official designation of "Area of Outstanding Natural Beauty" (AONB). The text elaborates on this theme:

> Travellers riding up the Snowdonia mountain railway may experience commu-
> nication problems. Your boss can't reach you. Even dogged telesales reps strug-
> gle. Damn those impenetrable mountain passes. Damn them. But the higher up
> you go the better the signal becomes. The view at the top is too good to keep
> to yourself. (http://icwales.icnetwork.co.uk/yourwales/tourism/tm_objectid=
> 14980940&method=full&siteid=50082name_page.html. Accessed September
> 24, 2005)

In the *Times* article we are treated to a righteous resentment directed at the stranded couple and group of management professionals who have unduly called out the mountain rescue service, reducing it to a "nannying and taxi service." This irresponsibility is seen to be reflected not only in the profligate use of the rescue services but also in the lack of respect directed at the natural environment itself: these urbanites have failed to prepare themselves properly for the potential rigours and routine uncertainties of the Lake District. Clearly, moral judgments are being cast on the behaviour of these individuals. Or to put it another way, a (not so) tacit contrast is being drawn between the comfortable, "soft" technonatural spatialities of the city, which are characterized by service, convenience, and immediacy, and the "hard" technonatural spatialities of the countryside, which are marked by unpredictability, self-sufficiency, preparedness, and patience. These two moral spatialities have of course often leaked into each other—city dwellers' unpreparedness for country walks is renowned (see Michael 2000); conversely, witness the recent sport of urban climbing (i.e., climbing urban structures, most dramatically the outside walls of skyscrapers). However, what is apparently especially irksome about the present incursion of the urban into the rural is that urban values and the expectations of urban convenience are being imported in such a bare-faced manner. The conduit or medium of this illegitimate transposition is—it is emphatically stated—the cellphone. This technology enables easy contact with rescue services, and as a result the culprits do not feel they have to bother to prepare properly for the potential rigours of the countryside. That is to say, the cellphone has cognitively corrupted these would-be walkers by extending a temptation they seem unable to resist. Indeed, this complaint implies that such behaviour entails an illegitimate ironization of rural space such that it also becomes "soft," full of convenience and service.

Now, it is no doubt possible to unpack this in terms of the rather tired arguments around technological versus social determinism (see Bijker 1995). For example, we could try to disentangle the way that the cellphone, by virtue of its capacity (ease of contactability), has transformed relationships (e.g., with particular services). Conversely, we could list the cultural or social prerequisites (e.g., consumerism, risk culture or society, individualization) that have facilitated this particular, unfortunate use of the cellphone. But this sort of analytic tack, along with the *Times* account (which I have recast in terms of contrasting but leaky technonatural spatialities), suffers from too much contrast. By this I mean that both are overly concerned, as it were, with unravelling ironies—or rather, with purifying ironic spatialities.

The dichotomy of spatialities collapses when we recall that the Lake District is not a "wilderness" in any simple sense, but a tourist destination of long standing. Fell walking is just one activity out of many (which include visiting gardens, mansions, museums, and horticultural shows;

boating; cycling; horseback riding; and taking guided tours around towns). In other words, the Lake District is a complex spatiality incorporating multiple technonatural relations. Indeed, contrary to the iconic status of the Lake District as the embodiment of the English Romantic wilderness, its landscape historically has been worked and reworked through a range technonatural relations (see, for example, Day 1996; MacNaghten and Urry 1998). In this context, cellphones are not simply artifacts that undermine the "wilderness" (as both a moral and a technonatural category) of the Lake District. Insofar as they mediate interactions that enable people to take advantage of what the Lake District has to offer, cellphones also help constitute it as a tourist destination, with all the convenience that comes to be associated with such destinations. What is enacted in the *Times* piece is the contest over the "purity" of the Lake District as an exemplar of "wilderness" and its "corruption" (urbanization, even?) as a tourist honeypot. However, the article goes on to note that Mountain Rescue services in the Lake District are funded by donations. This suggests that tourists are a substantial source of income for Mountain Rescue (which is nothing if not a socio-technical assemblage, after all). Here, then, we touch on an ironic technonatural spatiality that encompasses the complexities of the tourist gaze (Urry 1990): combined with the tourist ear and the tourist purse, the tourist gaze, in transgressing boundaries between the urban and the rural, is it would seem at once disparaged and welcomed.

When we turn to the Wales Tourist Board ad, the ironic character of technonatural spatiality is thrown into further relief. In that ad, it is not the ease of mobile communication per se that is bemoaned but—albeit ironically—its sporadic character. The reproduction of urban contactability is undermined by the interventions of those damned "impenetrable mountain passes." In other words, the dangers posed by the environment, including the terrain, relate not only to human bodies and capacities but also to mobile communications. This immediately brings into focus the fact that cellphones are part of a socio-technical assemblage that operates neither seamlessly nor universally. In other words, the cellphone's functionality is enabled by an electromagnetic medium that, ostensibly at least, obliterates distance, or rather turns qualitative space into quantitative space. Needless to say, such obliteration is itself mediated by the local base stations and masts that "support" that medium.

Furthermore, as we might also expect, these have become the foci of public concern. As Alex Law suggests, resistance to the siting of masts arguably reflects a desire to reassert local distinctiveness in the face of the "routine nomadic intimacy of cellphones [which] establishes place as an indifferent backdrop to being always 'on-call,' too absorbed in the 'busyness' of life to notice what is close at hand" (A. Law 2004, n.p.) — that is, the "erosion" of qualitative space. We shall return in due course to further interrogate this assemblage. In the meantime, drawing on Law and on the Wales Tourist Board ad, we can add another layer of meaning to the indignation expressed in the *Times* article. Cutting across the contrast between urban and rural technonatural spatialities is the (not uncommon) contrast between qualitative and quantitative space: the latter suggests paying attention, belonging, care; the former implies abstracted empty space that needs to be, one way or another, traversed as rapidly as possible (see Kern 2003 on the historicity of space). Qualitative and quantitative apprehensions of space apply to both urban and rural: cityscapes are no less valued as dwelling places even as they are abstracted. As we shall see below, if this contrast problematizes the one between urban and rural, in turn it too will be found to be problematic.

The Wales Tourist Board ad enacts yet another irony. In part its humour lies in the reversal of values. The mountain passes are negatively valued in that they serve to disrupt positively valued calls from the boss and "dogged telesales reps." Of course, it is the passes that are valued because of their beauty or magnificence; and it is the calls of boss and "dogged telesales reps" that are derogated because they interfere with such beauty or magnificence. Moreover, the mountain passes, because they break up cellphone signals, are further valued in that they serve to reinforce the sense of escape and of the exotic, of being away from the chores, pressures, and indeed inconveniences of everyday life. On one level, then, the irony of the advertisement helps de-ironize the technonatural space of Snowdonia: the contrast is drawn between the seeming isolation of Snowdonia and the hurly-burly of everyday urban life.

Yet notice that this engagement with the splendour (and convenient inconvenience) of Snowdonia is conducted from the relative comfort and convenience of a tourist train — a tourist train that is, not uncommonly, full of people. So the promise of isolation implicit in the ad is, ironically, realized in the midst of a touristic throng. On top of this irony lies another. The use of the cellphone can serve in the "doing" of a barrier between self and others on the commuter train (i.e., the management of public and private spaces); whereas on the tourist train, the place of the no longer functioning cellphone is taken up by Snowdon's majesty. The barrier between public and private can now be performed through the doing of intent absorption in the glories of Snowdon. At the same time, the view of Snowdon is all around, a common resource for the doing of privacy that ironically opens up the possibility of a sort of "shared privacy"—one that is partly mediated by the use of the camera facility on the cellphone to capture the sublime beauty of Snowdon. Here, insofar as the others' absorption is recognized, this can reinforce one's own absorption—something akin to what we might call the "collective sublime" (to adapt Nye's [1994] notion of technological sublime). This is convoluted still further with the recovery of the signal ("But the higher up you go the better the signal becomes. The view at the top is too good to keep to yourself"). Thus, one is able to share the view, presumably with those back in the city. At this point a publicly shared, privately experienced version of the sublime is shared privately with others at the other end of the cellphone while being witnessed by others on the train who are themselves presumably engaged in private cellphone conversations publicly enacted about a private-yet-publicly enacted experience of the sublime. And we might suspect that the mass cellphone calling is itself a topic of private cellphone and public local reflection. In any case, once again the divide between public and private seems dizzyingly to collapse.

In this section we have tentatively entered into the material–semiotic black hole that is the cellphone-in-the-countryside. In the process we have explored some of the complex ironies of the technonatural spatialities that mark the cellphone-in-the-countryside. Of course, no claim to exhaustiveness is staked. Rather, the aim has been to illustrate in some small way how the cellphone allows us to explore how rural technonatural spatialities can be characterized in terms of such dichotomies as rural–urban, public–private, qualitative–quantitative, and exotic–mundane—dichotomies that are at once purged and collapsed. In the next section we pursue this further by paying closer attention to some systemic dimensions of the cellphone-in-the-countryside—that is, its embeddedness in and embodiment of complex (and ironic) socio-technical assemblages and the "risks" associated with these.

Assembling Ironic Technonatural "Risks"

Up to this point our discussion has focused on the cellphone-in-the-countryside as a singularized object (this is itself a heterogeneous accomplishment, as it is for all technological artifacts). Yet this cellphone-in-the-countryside is a "constitutive product" (to coin another ironic term) of an expansive, variegated, and sometimes fractured socio-technical assemblage that spans, at a minimum, phone and related equipment manufacturers, a panoply of retailers and service providers, agencies of standardization, regulatory authorities, designers of infrastructure, medical bodies, and assessors of risk (and, needless to say, social scientists). In light of this character of the cellphone-in-the-countryside, we will now consider two additional ironic technonatural spatialities. As we shall see, these are crucially concerned with riskiness—or rather, with a variety of risks whose interrelations are, inevitably, ironic.[6] In what follows we consider two ways in which the cellphone-in-the-countryside assemblage can help reconfigure comportment by rendering aspects of the countryside more, or less, "visible." On the one hand, we consider the complex spatialities that emerge in relation to the potential use of GPS (Global Positioning System) enabled cellphones in the countryside; on the other, we explore the ways in which people might reorient around the "Teslar watch," given the seeming ubiquity (and dangers) of the electromagnetic waves that carry cellphone signals.

GPS and the Ironic Skeletalization of the Countryside

In recent years, cellphones have developed the capacity to process GPS information. Whether this capacity comes ready enabled (as in the Mio A701 GPS cellphone) or as an add-on (as in the TomTom Mobile 5 GPS for Nokia Smartphones), GPS can be used to establish the quickest or shortest routes for a range of modes of transportation, to provide 3D map views, and to accommodate customized points of interest. In relation to the cellphone-in-the-countryside, insofar as the GPS facility can inform users of local geographical features, it can—potentially at least— enable a process of "self-rescue," so to speak. In other words, GPS-enabled cellphones-in-the-countryside can become moralized and thereby become indispensable for the responsible walker or climber. Furthermore, to the extent that these phones mean that walkers have less need to call on local rescue services, they underscore the self-sufficiency of the walker. Of course, this enhanced self-sufficiency—or, to put it another way, this individualized, autonomized spatiality—is based on the dispersed spatiality of GPS, which is a thoroughgoingly heterogeneous and relational assemblage that famously incorporates some two dozen military satellites.

However, as things stand, it would appear that GPS-enabled cellphones have yet to attain the status of crucial equipment for the countryside. According to the British Mountaineering Council, "a GPS is not a substitute for use of a map and compass either in terms of accuracy or reliability and cannot be used as a sole navigational tool." Nevertheless, "a GPS system can offer useful additional information on which to base mountain navigation" (http://www.mountains-snowandrock.org.uk/Advice/advice_3.php, accessed August 18, 2006). Furthermore, as Lorimer and Lund (2003) note, hill walkers have a complex relation to GPS: "It is variously understood as a marker of increasingly professionalised practice, a mechanism to reduce personal risk, a dangerous gadget likely to encourage acts of irresponsibility and an intrusive presence in the hillwalking experience" (2003, 141). This complexity will doubtless be further emphasized by the circulation of danger narratives in which too much reliance on GPS led to easily avoidable dangers (the pensioners' coach guided into, and stuck fast in, an unsuitably narrow country lane; the car that was guided to the edge of a precipice and a hundred-foot drop, etc.).

So the capacity of GPS-enabled cellphones to reconfigure the responsibilization of the walker or climber remains, not unexpectedly, underdetermined. Yet GPS as a potential tool for navigation (and as a medium of responsibilization) shares some key features with the map—features that contribute to another ironic technonatural spatialization. Both GPS and the map (in Britain, most often the Ordnance Survey map) abstract standardized details from the environment—in this, there is a systematic *neglect* of information—in order to produce a "useful" cartographic representation of the terrain and its specific features (see, for example, Marsden and Smith 2005). In a sense, there is a "skeletalization" of the countryside whereby detail is stripped in order better to guide the walker. However, this skeletalization can only be operationalized if at least two conditions are in place. First, there must be some form of trust in the skeletal representations depicted by the GPS map—that over that horizon there will be a village, that beyond that field there is a pub, that around that corner the gradient reduces. The self-sufficiency enabled by skeletalization is in this way tacitly grounded not only in the relationality to the communication system, but also in those expert bodies that skeletalize the countryside (e.g., the Ordnance Survey, "the national mapping agency of Great Britain"). The second condition is that skeletalization is only workable if "flesh" is added.[7] What I mean by this is that the embodiment of the walker and the use of a range of senses is necessary to implement the directions/quasi-instructions, and enact the abstractions, performed by the GPS.[8] To negotiate the terrain, the walker must take note of the specificities of local spatiotemporal conditions: here and now, this particular body is tired, that lane is full of sheep, this river is about to burst its banks, that slope is too unstable. The fleshly reading of the countryside (a space of enactment and incorporation, as opposed to inscription and representation) thus mediates the spatial mediation of GPS—ironically, then, the skeleton hangs on the flesh.

However, this irony folds into another. We should not oppose these two registers too starkly (see Lorimer and Lund 2003). The cartographic GPS as an expertly reduced representation can, of course, also enhance the embodied engagement with the countryside. To know that one is safe can open up a space for the romantic experience of the countryside: one doesn't stumble on the sublime, one needs to accomplish it, heterogeneously (see Michael 2000). To replenish the senses in nature after the onslaughts of the urban lifestyle (see Thomas 1984), one first needs to get out into "nature"; and GPS, associated with a reduced sensorium as it is, can aid in this. Furthermore, insofar as such knowledge provides otherwise unavailable information—depicts an aerial overview, points to places of interest, supplements the tourist gaze with historico-ecological knowledge—the intercorporeality of the experience can be "deepened." To walk along a Roman road as opposed to a path, to hike by an oxbow lake as opposed to an oversized puddle, to move through an ecological community as opposed to being bitten by midges—these expert accounts contribute another dimension to the enactment of the cellphone-in-the-countryside. But it should be added that if GPS supplies additional information, this does not always render the experience more "profound"—it can be a distraction too.[9]

In this section we have briefly explored some ironic technonatural spatialities mediated by GPS-enabled cellphones. We have seen how these at once "moralize" and "technicize" the countryside in highly complex and involved ways. As GPS retechnicizes technonatural spatiality (i.e., skeletalizes it), it also remoralizes it (how should one comport oneself as a responsible walker, or tourist?). Conversely, the existing technical practices of walkers and tourists also serve in the moralization of GPS (is the use of GPS a "good thing"?). In the process, of course, such moralization — technicization of technonatural spatialities also pulls in, for example, the body, identity, social relations[10] — or rather, serves to underline how technonatural spatialities are comprised of these entities and of relations as well.

The Ubiquity of Electromagnetic Risk: The Ironic Haven of Re-De-Nature

As mentioned above, one view of the countryside is as a place of replenishment, recuperation, and reinvigoration. Yet the countryside (certainly in Britain) is also seen to be pervaded with all manner of risks, from those associated with the systems that transport people into the countryside (traffic accidents, train breakdowns, air travel pollution), through the multiple residues of industrialization (pesticides, fallout), to the "natural" dangers of sunlight[11] and mutating microbes. In the case of the cellphone-in-the-countryside, additional risks can become apparent. Most obviously, there are the possible risks to the brain posed by the phone's transmission of radio waves. As is well known, there is still much controversy over what exactly these risks might be.[12] Furthermore, there are the risks that might be posed by the base stations that are used to transmit radio waves to cellphones. Unsurprisingly, controversy surrounds this issue too.[13] Even though the likely exposure to radio waves is somewhat diminished in rural areas given that base stations are set farther apart, the general point is that this can be seen as another form of "pollution" and as another source of "artificial risk" in a context where the countryside is regarded as "natural." What I want to consider briefly is one possible response to this risk — the wearing of a Teslar watch. The objective here is to examine how such reactions (and this is one out of several possible mitigating moves) serve in the supposed re-establishment of an individualized natural spatiality — or rather a renaturing of a denatured natural space that is, ironically, profoundly technoscientific — that is, technonatural. This — what we might call "re-de-nature" — is further implicated in other cultural and social ironies.

For people who feel that their bodies are at risk from chronic electromagnetic exposure (e.g., from base stations, as well as from many other sources), Teslar watches have been developed that, it is claimed, draw on "scalar energy" to magnify "the strength of the biofield, thus protecting the body from destructive electromagnetism" (see Bioenergyfields

Foundation, http://www.bioenergyfields.org/index.asp). The key component, the Teslar chip, "emits a frequency of 7.83 Hertz known as the
Schumann Resonance. This is the frequency that the entire Earth's mag-
netosphere resonates at. When we are in natural environments away from
big cities, this is the predominant frequency which our body feels. It has a
profound 'healing' effect. When we go back to the city, there are many
other frequencies that override this resonance putting our body into a
more 'chaotic' state of being" (see http://www.toolsforwellness.com/teslar
-watches. html). Thus, the watch recreates the "natural environment" in the
urban setting, a "natural environment" that is, in and of itself, "healing."

Crucial to this account is the contrast between the countryside and the
city, where the former is romantically depicted as the "natural environment," whose healing powers are ironically uncovered and marshalled
through technoscience. There are further ironies to this contrast: the
Teslar watch is a product of the urban that recreates the natural; the natural in this case entails a mobile, individualized spatiality (the bodily space
of the watch) that cuts a swath through the unnaturalness of its environs—
natural nature becomes not an extended context, but a cocoon housing
the person—a personalized haven that is enabled by an extensive socio-
technical assemblage; nature has been denatured (even the "natural environment" is risky, given that it cannot escape the pollution of electromagnetic waves), then renatured through the Teslar watch—people now
occupy a re-de-nature.

However, controversy attaches to the Teslar watch. The underlying scientific principles and assumptions have, almost inevitably, been questioned (see, for example, http://www.wired.com/news/gizmos/0,1452,
60183,00.html). There is, in other words, a charge of quackery directed at
the Teslar watch. Furthermore, celebrities (e.g., Oprah Winfrey) who have
endorsed the watch have been ridiculed for their gullibility and their willingness to advocate a spurious technology on the basis of dubious expertise and suspect experience (see, for example, http://www.gizmodo.com/
gadgets/gadgets/teslar-watch-oozes-wellbeing-into-oprah-171061.php).
Rather than "inoculating" the Teslar watch against ridicule (see Michael
1997), such associations with celebrity are seen by some to render the
watch even more problematic. Once again, in the making of technonatural spatiality (in this case re-de-nature), trust and credibility come into
play. Arguably, in the instance of the Teslar watch, the believability of the
Teslar's re-de-natured cocoon lies less with the credibility of technoscience

and more with the dynamics of celebrity culture. To put it another way, the technonatural spatialities of the Teslar watch's electromagnetic cocoon are ironically mapped onto the complex spatialities of celebrity (which, as we have seen, are in no small part characterized by the "ridiculous").

Concluding Remarks

It goes without saying that in this chapter I have explored a rather circumscribed version of "technonatures": the cellphone-in-the-countryside. Even my account of this has been relatively limited and perhaps a little peculiar: a more obvious analysis might have looked at, perhaps, the way in which cellphones are instruments of environmental intervention, serving in the coordination of direct action. However, I have been more interested in considering the cellphone as a topic rather than a resource — a topic for illustrating some small aspect of contemporary technonatures. In developing and deploying the concept of "ironic spatiality," I hope I have illuminated some of the complexities of technonatures. At the very least, we have encountered the way in which the icons of the countryside such as the Lake District and Snowdon are saturated with ironies in which the urban and the rural, the qualitative and the quantitative, the private and the public, the safe and the risky, interweave. Along the way, other notions such as skeletalization and re-de-nature have been suggested as a means to come to grips with some of the political ironies of technonature, not least as these implicate matters of trust and credibility, and conversely (corporeal) skepticism and ridicule. One general lesson is that the cultural and political edifices that are the Lake District and Snowdon (and these are merely illustrative of many other such edifices across the world) are built on shifting technonatural sands.

In other words, in tracing how several technonatural orderings coexist in the "same" instance of the "cellphone-in-the-countryside," we can perhaps broadly suggest that any politics might benefit from a sensibility that encompasses such ironies. In the place of an environmental politics grounded in a purified version of nature, we need a politics that addresses the complex and dynamic impurifications of technonatures. Of course this is much easier said than done, not least insofar as the complex and dynamic impurifications of technonatures are as liable to alienate and disenchant as to inspire. This reflects the history of Western conceptions of the natural environment and what is found enchanting therein (notions of wilderness, standards of beauty). Yet such enchantment is

founded on systematic forgetfulness. The "event" of an enchanted relation with the natural environment is one that is comprised of the sort of
ironies we have explored in this chapter. At minimum, the nature we
value and are enchanted by, and to which we are attached, is always
already a mediated one: for example, various modes of representation
(photographic, fictional, cartographic, classificatory) and of transport
(trains, cars, walking boots) have had to be in place for such relations
with the natural environment to be "do-able."

If this chapter has gone a little way to unpacking how such enchantment and attachment are in actuality rooted in the shifting sands of
"ironic technonatural spatialities," this still leaves open the issue of how,
in a world of dynamic complexities, enchantment and attachment are to
be re-established. One move is to begin to seek these in dynamic complexities—to recognize that the sorts of ironies described here have their
own appeal (not least aesthetic) that initiates and sustains environmental
political practice. But then again, perhaps this is the wrong move—perhaps enchantment and attachment are part of the problem. After all, both
these terms imply a pre-existing subject that is "enchanted by" and
"attached to" (whether to an iconic environmental edifice or to a dynamic
technonatural irony)—a subject that lacks the fluidity or processuality to
do justice to the dynamic complexities evoked by "ironic technonatural
spatialities." Instead—and perhaps this doesn't warrant the name "politics"—what is needed is a "politics" that encompasses the fact that political actors emerge from (and are mediators of) these very technonatural
complexities.

Notes

1 The present chapter uses the term "cellphone," though in the British context
  "mobile phone" is more common. Some of the data presented here refer to
  the latter, but cell and mobile should be regarded as interchangeable for
  present purposes.
2 On this score, "the-cellphone-in-the-countryside" serves as a black hole, as
  Haraway (1994) would call it: dense knots or "concrescences" (Whitehead
  1978 [1929]) in which the components of these dichotomies are mutually
  entangled.
3 While the present emphasis is on technonatural spatiality/spatialization,
  it is presupposed that such processes cannot be divorced from temporal-
  ity/temporalization—whether at the "macro" level (e.g., the historicity of

spatiality—see, for example, Lefebvre 1974) or the "micro" level (e.g., the local "topologization" of time—see, for example, Serres and Latour 1995). See also White and Wilbert (2006).

4 Here I draw on the argument of Akrich and Latour (1992, 259) who state: "the word 'meaning' is taken in its original nontextual and nonlinguistic interpretation: how a privileged trajectory is built, out of an indefinite number of possibilities."

5 Of course, this version of the urban social has recently been overhauled as social scientists—most especially geographers—have charted the complex presence of "nature" in the city, not least that of animals (e.g., Wolch and Emel 1998).

6 This attendance on "risk" could be seen as a contribution to recent discussions of "risk society" (e.g., Adam, Beck, and Van Loon 2000). However, I am rather more interested in the ways that the-cellphone-in-the-countryside mediates a range of concerns that no doubt entail something like "risk," but also "social anxieties," "cultural ambivalences," and ironic spatialities that cannot be readily distilled into a term such as "risk."

7 The terms skeletalization and flesh strike me as particularly apt in this instance because they echo Merleau-Ponty's (1968) use of the term "flesh" to connote the way that the environment "perceives" humans and embodies them in its processes. Here we are touching on the mutual embodiment of walkers and landscape (see Michael 2000).

8 This terminology is another way of portraying the processes addressed through the dichotomy of quantitative and qualitative spatialities.

9 On this score, any communication technology needs to be considered in relation to the multiplicity of other communication practices and technologies that circulate within a particular community, or assemblage—that is, a "communicative ecology" (see, for example, Horst and Miller 2006).

10 Additionally, we can note that GPS has been lauded (partly ironically) as a saviour of spousal relationships: the arguments so typical of driving-and-navigating can, apparently, be diffused or settled by the GPS's more authoritative voice.

11 Regarding sunlight, this obviously has a complex history and has been seen a boon to health as well as a risk (see Carter 2007).

12 Thus, while there is now a standardized methodology for measuring the absorption of radio waves in the head for handset models—the specific absorption rate (SAR)—there are several complicating variables—notably, the distance of the phone from the head and the frequency and duration of calls (see, for example, the Health Promotion Agency's Web page on Mobile Telephony and Health: http://www.hpa.org.uk/radiation/understand/information_sheets/mobile_telephony/mobile_phones.htm; accessed September 8, 2006).

13 See, for example, the recent exchange between Hutter and colleagues (2006) and Coggon (2006).

Works Cited

Adam, B., U. Beck, and J. Van Loon, eds. 2000. *The Risk Society and Beyond.* London: Sage.

Akrich, M., and B. Latour. 1992. "A Summary of a Convenient Vocabulary for the Semiotics of Human and Nonhuman Assemblies." In *Shaping Technology/Building Society*, ed. W.E. Bijker and J. Law: Cambridge, MA: MIT Press.

Amin, A., and N. Thrift. 2002. *Cities: Reimagining the Urban.* Cambridge: Polity.

Bennett, T. 2004. "The Invention of the Modern Cultural Fact: A Critique of the Critique of Everyday Life." In *Contemporary Culture and Everyday Life,* ed. E. Silva and T. Bennett. Durham: sociologypress.

Bijker, W.E. 1995. "Sociohistorical Technology Studies." In *Handbook of Science and Technology Studies*, ed. S. Jasanoff, G.E. Markle, J.C. Peterson, and T. Pinch. Thousand Oaks, CA: Sage.

Bull, M. 2000. *Sounding Out the City.* Oxford: Berg.

Carter, S. 2007. *Rise and Shine: Sunlight, Technology, and Health.* Oxford: Berg.

Coggon, D. 2006. "Health Risks from Cell Phone Base Stations." *Occupational and Environmental Medicine* 63: 298–99.

Cooper, G., N. Green, G.M. Murtagh, and R. Harper. 2002. "Mobile Society? Technology, Distance, and Presence." In *Virtual Society? Technology, Cyberbole, Reality,* ed. S. Woolgar. Oxford: Oxford University Press.

Day, A. 1996. *Romanticism.* London: Routledge.

de Certeau, M. 1984. *The Practice of Everyday Life.* Berkeley: University of California Press.

Gardiner, M. 2000. *Critiques of Everyday Life.* London and New York: Routledge.

Haddon, L. 2004. *Information and Communication Technologies in Everyday Life.* Oxford: Berg.

Haraway, D. 1994. "A Game of Cat's Cradle: Science Studies, Feminist Theory, Cultural Studies." *Configurations* 2: 59–71.

Highmore, B. 2002. *Everyday Life and Cultural Theory: An Introduction.* London: Routledge.

Horst, H.A., and D. Miller. 2006. *The Cell Phone: An Anthropology of Communication.* Oxford: Berg.

Hutter, H.-P., H. Moshammer, P. Wallner, and M. Kundi. 2006. "Subjective Symptoms, Sleeping Problems, and Cognitive Performance in Subjects Living near Mobile Phone Base Stations." *Occupational and Environmental Medicine* 63: 307–13.

Katz, J.E., and S. Sugiyama. 2005. "Mobile Phones as Fashion Statements: The Co-creation of Mobile Communication's Public Meaning." In *Mobile Communications: Re-negotiation of the Social,* ed. R. Ling, and P. Pederson. New York: Springer.

Kern, S. 2003. *The Culture of Time and Space, 1880–1918.* Cambridge, MA: Harvard University Press.

Latour, B. 2004. *Politics of Nature.* Cambridge, MA: Harvard University Press.

Law, A. 2004. "The Social Geometry of Mobile Telephony." *Razón y Palabra* 24 [online journal]. http://www.razonypalabra.org.mx/anteriores/n42/alaw. html#au.

Law, J. 2004. "And If the Global Were Small and Noncoherent? Method, Complexity, and the Baroque." *Environment and Planning D: Society and Space* 22: 13–26.

Lefebvre, H. 1974. *The Production of Space.* Oxford: Blackwell.

———. 1968. *Everyday Life in the Modern World.* London: Allen Lane.

Lorimer, H., and K. Lund. 2003. "Performing Facts: Finding a Way over Scotland's Mountains." *Sociological Review* 51, no. S2: 130–44.

MacNaghten, P., and J. Urry. 1998. *Contested Nature.* London: Sage.

McGuigan, J. 2004. "Toward a Sociology of the Mobile Phone." *Human Technology* 1: 45–57.

Marsden, B., and C. Smith. 2005. *Engineering Empires: A Cultural History of Technology in Nineteenth-Century Britain.* Basingstoke: Palgrave Macmillan.

Merleau-Ponty, M. 1968. *The Visible and the Invisible.* Evanston, IL: Northwestern University Press.

Michael, M. 2006. *Technoscience and Everyday Life: The Complex Simplicities of the Mundane.* Maidenhead: Open University Press/McGraw-Hill.

———. 2000. *Reconnecting Culture, Technology, and Nature: From Society to Heterogeneity.* London: Routledge.

———. 1997. Inoculating Gadgets against Ridicule. *Science as Culture* 6, no. 2: 167–93.

Muecke, D.C. 1969. *The Compass of Irony.* London: Methuen.

Nye, D.E. 1994. *American Technological Sublime.* Cambridge, MA: MIT Press.

Persson, A. 2001. "Intimacy among Strangers: On Mobile Telephone Calls in Public Places." *Journal of Mundane Behavior* 2, no. 3. http://www.mun danebehavior.org/issues/v2n3/persson.htm.

Poster, M. 2002. "Everyday (Virtual) Life." *New Literary History* 33: 743–60.

Serres, M., and B. Latour. 1995. *Conversations on Science, Culture, and Time.* Ann Arbor: University of Michigan Press.

Shields, R. 1999. *Lefebvre, Love, and Struggle: Spatial Dialectics.* London: Routledge.

Thomas, K. 1984. *Man and the Natural World.* Harmondsworth: Penguin.

Urry, J. 1990. *The Tourist Gaze.* London: Sage.

Whatmore, S. 2002. *Hybrid Geographies.* London: Sage.

White, D., and C. Wilbert. 2006. "Introduction: Technonatural Time-Space." *Science as Culture* 15, no. 2: 95–104.

Whitehead, A.N. 1978 [1929]. *Process and Reality: An Essay in Cosmology.* New York: Free Press.

Wolch, J., and J. Emel, eds. 1998. *Animal Geographies: Place, Politics, and Identity in the Nature–Culture Borderlands.* London: Verso.

Chapter Four

Living Cities: Toward a Politics of Conviviality
Steve Hinchliffe and Sarah Whatmore

Against the cartographic opposition between cities and natures in modern Western societies, the idea of urban ecology has seemed little more than a contradiction in terms. However, the spaces and species that have been erased from urban visions and values now find themselves the subject of a "greening" of urban policy that has gathered some momentum in the United Kingdom on at least two fronts. First, urban biodiversity is *starting* to be accorded the kind of conservation significance once reserved for rural and sparsely populated regions. So much so that the distinction between greenbelt and brownfield land is no longer an automatic marker of ecological value. In this, scientific energies are being newly invested in the importance of so-called "recombinant ecology" (Barker 2000). This refers to the biological communities assembled through the dense comings and goings of urban life, rather than the discrete and undisturbed relations among particular species and habitats that are the staple of conservation biology. Urban wildlife groups, amateur naturalists, and voluntary organizations, no less than the highly visible animals and plants that make their way in and through cities, have been key players in this realignment of urban spaces and conservation concerns. Second, there is a growing sense in the urban-policy community of the importance of this "recombinant ecology" to what makes cities livable and to the attachments of civic identity and association. Critical here is the extent to which this ecological fabric is constituted as a public good or urban commons, including leisure spaces such as parks and allotments;

feral spaces such as abandoned railway sidings and derelict land; and remnant spaces such as waterways and woodlands. This gathering of energies has found policy expression in new political investment in what has become known as the "urban green."

The research we draw on in this chapter is an intervention in this "greening" process and, as such, stretches the terms of policy investments in the "urban green." In particular, through the notion of "living cities" we want to articulate some significant challenges to the styles and practices of analysis institutionalized in scientific and policy procedures that characteristically presume and reinforce spatial divisions between civic and wild, town and country, human and non-human. To this end we are working toward a conception of living cities that resists this familiar architecture of urban analysis in at least three ways: by exploring (a) how cities are inhabited with and against the grain of expert designs—including those of capital, state, science, and planning, (b) how urban inhabitants are heterogeneous, made up of multiple differences mobilized through human and non-human becomings, and (c) how urban livability involves civic associations and attachments forged in and through more-than-human relations.

In the first part of the chapter we expand on this list of challenges to the architecture of urban analysis, drawing on research on the spatialities of a number of urban inhabitants. We open up some of the possibilities to be found in reworking the city as a living environment and follow up some of the consequences these have for the ways in which people and others map the city. The examples are drawn from one city in particular—Birmingham—and from work with urban gardeners, conservationists, and other activists.

In the second part of the chapter we expand on this commitment in order to sketch a politics that attends to the "leaks" and "overspills" of primary forms of political alignment and organization and that disrupts the analytical and policy framing of what, and how, "politics" are performed (Rose 2000; Tully 1999). We do so by outlining three components of what we take to be intrinsic ingredients of a *politics of conviviality*, involving humans and non-humans. Our attempt to formulate such a political project here draws on our research involvements in the activities of living cities.

Living Cities

In this section we move through the three ways in which living cities challenge the familiar architecture of urban analysis in turn, drawing on engagements with urban ecologies in Birmingham.

Living With and Against Urban Designs

We can start close to the centre of Birmingham, at the tallest building, the telecommunications tower. In among this thoroughly built and planned environment live raptors — in this case peregrine falcons. Contrary to expectations, as this environment continues to be evacuated in policy and scientific accounts of all but human life, they have made themselves at home here and have found enough food to stay year round and breed (Whatmore and Hinchliffe 2003). In much urban theory, the only active participants in this urban world are the people who busy themselves beneath the tower and who reach out from the tower to all corners of the world through its satellite dishes, computer networks, and other devices of connectivity. The materiality of cities is presented by turns as passive and inert; concrete is treated as the paradigmatically malleable material of the civic realm. Yet the peregrines suggest that there is more to city living than technology and culture — or, more tellingly, more to technology and culture than human design.

It's not just people, peregrine, and pigeon that cohabit the designs of urban space. A few metres from the tower one encounters part of Birmingham's canal network. This section has recently been dammed and dredged to facilitate the redevelopment of canalside buildings. Ten species of fish have been recorded in this unpromising-looking stretch of water. There's also now a very real possibility that otters (*Lutra lutra*) will soon be moving through the city's canals. They're not quite in the city centre yet, but only a few miles away there are complaints from anglers that fish stocks are dwindling in urban canals, rivers, and ponds, and otters have been caught on CCTV passing under road bridges and through culverts. It's widely agreed that it won't be long before otters pass this way. Meanwhile, across the road from the tower there is a derelict building and a piece of land currently used as a makeshift car park. Hereabouts lie a number of ancestral breeding sites for the black redstart (*Phoenicurus ochruros*), which is Britain's rarest bird.

We could go on with this catalogue. We haven't as yet mentioned the myriad micro-organisms, plants, and insects that make cities living spaces, nor have we opened up the question of what is and isn't living in cities. Our point here has only been to underline that cities are inhabited with and against the grain of expert design. But we can and will go further. For the examples we have mobilized so far speak only of cohabitation. On their own they do little to unsettle the bounded cartographies of technology and nature. If they do anything, they simply suggest that those cartographies exist on a different scale than that supposed by urban planners. We now want to argue that just as city living is not unaffected by this multiplicity and indefinable variety of inhabitations, so inhabitants are not unaffected by city living. Indeed, we want to suggest that non-humans don't just exist in cities, precariously clinging to the towers and edifices of modernity: they are shaped by and have the potential to shape their urban relations. Nor do we see these inhabitants as a threat to modernity as — perhaps and in more prosaic terms — the owners of the Telecom tower may view the peregrine. Rather, we would like to suggest that the demography of the city, its populace of human and the non-human inhabitants, unsettles the geography of modernity and its forebears. This is, then, an instance of what Latour would call an amodern engagement (Latour 1993).

Urban Inhabitants

If cities are inhabited with and against the grain of urban design, such inhabitation involves more than living with the city. It involves ecologies becoming urban and cities becoming ecological. Here we can draw on ethology — the science of evolution, or, more resonantly in the terms of our analysis, the science of becoming. We are not alone in suggesting that current reinvestments in the practice of this science are potentially creative (see, for example, Ansell Pearson 1999; Thrift 2000). Lives may not simply act out an internal script, checked only by external conditions. They may rather be material enfoldings of complex topologies of living and non-living entities (Whatmore 2002). In other words, we might see urban inhabitants as more-than-human; more than animal; more-than-plant; and so on. They are complex assemblages, mutually affecting and affected by their fields of becoming. This tack implies that urban ecologies are urban in more ways than their familiar guise as a set of obstacles or opportunities contrived by human technologies and human presence. They are more mixed up than that, as another example serves to illustrate.

Three kilometres south of Birmingham's Telecom tower is a large area of land that is variously regarded by landowners and would-be developers as a brownfield or derelict site ripe for development. Through it runs a small stream known locally as the Bourn Brook, which is bordered and crossed by canals and railways and which links up to a complex patchwork and network of urban land forms. Ecologists think that this is suitable habitat for the water vole (*Arvicola terrestris*) and have found evidence of this small mammal even as it becomes increasingly rare in rural Britain. Indeed, outside a few urban areas, water vole populations have declined dramatically, leading some to suggest that water voles may be the next mammalian extinction in the British Isles. For conventional ecology, with its implicit and sometimes explicit hierarchy of spaces from the pure to the despoiled, this relative success of the urban water vole is a surprise. In this schema, cities are the last places to find refuge and the first places from which nature has fled. Even more puzzling is the possible cohabitation on this site of water voles and a number of predators and competitors, especially the brown rat (*Rattus norvegicus*). Perhaps, just perhaps, the entangled and multiple topologies of "the site," which has been shaped by numerous human and non-human activities (from river courses to quarries, from willow trees to canal networks), affords the conditions of possibility for water voles to inhabit these spaces against all odds (for more detail see Hinchliffe et al. 2005).

Water voles aren't the only mammal to be doing well in cities or to be living and passing through this site (see Wyatt 2003 for an attempt to map Birmingham's mammal populations). Badgers (*Meles meles*) in particular have been successful in Birmingham, and there is growing evidence that their foraging ranges and their group sizes (in short, their territoriality) are very different in these urban habitats than ecologists are used to observing "in the wild." Indeed, badger enthusiasts and naturalists are starting to accumulate observations which suggest that badgers are less likely to define territorial boundaries in urban areas and have larger foraging ranges than their rural counterparts (Wyatt 2003). Their paths through the city seem to extend along linear features (especially railways and canals) — a mapping that is more attentive to movement than to fixity and that articulates the spatialities of networking rather than those of territory in the Euclidean sense.

As our tentative vocabulary has sought to suggest, these vignettes of animal inhabitation fall short of being established "matters of fact" and mark a neglected research topic. In the introduction to this chapter we alluded to the ways in which urban ecologies are often treated as "poor cousins" in the production of ecological knowledge. Established forms of knowledge practice have often eschewed ecological research in cities, observing the logic of cultural divisions between the social and the natural and their respective mappings onto urban and rural spaces (for exceptions see Hough 1995; Mabey 1999; Matthews 2002; Wheater 1999). But this also speaks of something else, something that is beyond this antiurban bias in expert ecological knowledge practices. It articulates an intensified indeterminacy, one that marks all ecologies but that is perhaps simply more apparent in urban habitats. The ecologies glimpsed here are inhabited and enacted by beings-in-relation that may be better thought of as becomings that enfold human and non-human mappings. Rather than being matters of fact, these vignettes of urban ecologies are better appreciated in terms of what Latour (2004) would call matters of controversy—that is, as matters entangled in all manner of ways and with all manner of things. Entanglements that make life more interesting than a series of smooth entities that furnish the world as matters of fact, once and for all. As entangled matters of controversy, they are lived realities that can and do demand responses and entail all sorts of obligations (see also Serres 1995). It is to the style or technologies of these responses and obligations that we now turn.

Making Cities Livable

Our third challenge to the architecture of urban analysis revolves around civic associations and attachments. This challenge may be more effective if it is posed as a question: How can civic attachments and associations adapt to the hectic comings and goings of living cities? We can suggest what is at stake by returning to the Bourn Brook. It is not at all obvious that water voles live here. Indeed, one survey of the site (independently commissioned as part of an environmental assessment on behalf of the developers) found no evidence of water vole activity and questioned the presence of a whole suite of species that local activists and Birmingham and Black Country Wildlife Trust ecologists had recorded (Babtie 2001). Part of the problem here is the difficulty of reading any landscape for the life signs of creatures that are often subtle and transient (Hinchliffe et al. 2005). But there is also the more complex issue of whether the question of something's being present or not is as black and white as it seems. Indeed, the mobilities of heterogeneous urban inhabitants are not always easy to map. Their spatial and temporal practices may well be more intermittent, durable, and/or fleeting than is allowed for by conventional technologies of representation. This in turn raises questions about how best to form the kinds of associations that befit living cities.

Some of the answers to these questions are well known or, we might more suggestively say, are well practised. Maps of species presence give way, *in practice*, to more nuanced understandings of urban ecologies (see Latour 2004 for a similar call to attend to the practice of political ecology rather than its theorization). So rather than seeking the political short-cuts that seem possible through legal structures and straightforward presence, there are more open experimentations in how sites such as the Bourn Brook can be enacted in ways that are good for all potential inhabitants. This experimentation might at first seem fanciful compared to the more secure-sounding politics of representation, but to those who are experienced in ecological practice it is recognizable as a familiar pragmatic — and sometimes covert — means of "doing" urban ecology. This way of proceeding may simply need more ready experimentation in order to head off the potentially damaging consequences of relying politically on species or other entities being present in the landscape. When attention is turned instead to "making present," the work of fabricating or enacting the possibilities for living cities comes into focus.

Such experimentation is already evident in a variety of practices, which are misleadingly labelled as "expert" and "lay" but which are more tellingly cast as living "inexpertly" in the manner of the Deleuzian formulation of the impossibility of expertise. What we mean to say here is that all of these experimental activities involve elements of not knowing, of the unknown and the unknowable. They *are* informed through experience, but they are also likely to throw up surprises and to generate new configurations. This reformulation of ecological attachments as inexpert experiments can be illustrated by two examples from the species and spaces already mentioned. The first involves the black redstart, whose inhabitation of the city is not straightforwardly a matter of their presence in the city. Black redstarts are among the rarest birds in the British Isles, so it is very difficult to find them. They thrive in thermophillic environments, characteristic of shorelines and derelict land where ruderal plant species attract invertebrates, on which the birds feed. Such conditions were to be found most readily during the Second World War and in the 1970s and 1980s (courtesy of bombing raids and economic decline)

(Davis 2002). At that time there were something like fifteen breeding pairs in the city. Since then, ancestral breeding sites have been turned from canalside industrial dereliction to landscapes for human consumption. The loss of suitable habitat has very possibly contributed to a decline in Birmingham's black redstart population, which is now thought to be something like two to four breeding pairs (which is still 5 to 10 percent of the British population). The low numbers, along with generic difficulties attendant to bird surveys in cities, including noise levels, uneven topography, poor access to sites, and so on, mean that their presence in the landscape is difficult to ascertain in practical terms.

If presence can be demonstrated, protection measures for black redstarts in urban centres like Birmingham and London are quite strong. The bird's national rarity endows it with so much ecological importance in the traditional value system of established conservation that it receives as much legal protection as is currently attainable. It enjoys robust legal protection, including designation as a fully protected species on Schedule I of the Wildlife and Countryside Act 1981. That legislation provides protection for the birds, their eggs, and their nestlings from killing or injury. It also covers the destruction of their nests and any intentional disturbance while they are building or attending to their nests. All of this protection is fine if presence is recorded. However, as with the water voles—who are as yet not so well protected by legal measures—there is the risk that disputed presence can make legal protection impotent. On the face of it, this seems sensible—legal protection of course requires some proof that there is something to protect. This is a pragmatic spatial mapping that can deal with all manner of ecological differences—as long as those differences are easily marked in terms of presence or absence. However, as we have already mentioned, non-human mappings take many forms, and in this case the vagaries of survey techniques and the "at best" flickering presence of inhabitants like black redstarts and water voles in the urban landscape point to other time/spaces. It would seem, then, that the legal topology of presence and absence is at risk of becoming a hit-and-miss affair in the business of conservation.

In the case of the black redstart, and in *practice*, a procedure has been worked out to cope with these complex spatial and temporal habits. A working category has emerged called "likely presence"—a term that is less of a hostage to the problem of securing presence that we discussed earlier and that ecologists in Birmingham are now using in their dealings with would-be developers in the city centre. "Likely presence" is enough to require that a survey be carried out. Given the costs of surveys, developers are now more inclined to opt for mitigation without establishing once and for all (if such a thing were possible) the presence or absence of black redstarts. It can turn out to be cheaper to act as if there was presence, and/or to be seen to be advancing the potential for nesting, and/or to experiment with different forms of landscaping and building, than to go through the drawn-out surveying and legal procedures that may or may not determine presence or absence. Thus, developers in Birmingham and London are now installing relatively intricate nesting boxes (black redstarts, like many other birds, prefer particular built forms, which nest boxes need to mimic in some fashion). They are also reusing hard-core and other "waste" as low-nutrient substrate for plantings of ruderal species; and increasingly, they are constructing green roofs. All of this work is experimental; its aim is to generate suitable habitat and at the same time reduce costs.

It is of course easy to be cynical here—to cite these developments as nothing more than minor concessions to a political lobby group that can do little in urban settings other than argue for mitigation measures. While it is certainly important to keep these measures in perspective, this should not obscure the point that a shift is occurring here away from a concern with the clearly present toward one of experimentation in urban ecologies, and away from a closed politics of ecological states of nature toward a more open politics of things, of living as others. Our second example takes this point a little further.

This reconfiguration of ecology, away from statements of fact toward engagements with possibilities, shifts the status and location of expertise. Rather than unveiling the truth of the ecology at hand, there is a turn toward those involved in the co-fabrication of living cities. There is a re-distribution of expertise, or a redefinition of expertise so that it includes lay engagements with place—gardeners as well as horticulturalists, amateur enthusiasts as well as professional ecologists. So, for example, engagements with a place on a day-to-day basis, or through less frequent but recurrent visits, can generate a sensibility about, or intimacy with, ecologies of place. Increasingly, this kind of intimate knowledge is being harnessed to fill the gaps in the time/space technologies of formal ecological knowledge practices, such as quadrants, grid squares, and fieldwork periods. Thus, city residents familiar with a particular site can act as the eyes and ears of ecological record centres and in so doing can add much needed longitudinal data to site records. Civic attachments and associations are also likely to take many other forms, forms that will produce different kinds of knowledge. From the routines of walking the dog or working an allotment to planting a tree or constructing a pond, all of these activities involve or enfold people and myriad living and non-living things. Enfolding lives in the way we are suggesting here is not so dependent on co-presence. More complex times and spaces are part of the ways in which these actions are understood and of the ways in which they evolve in practice. We would call all these kinds of attachments—including those that are formally part of ecological expertise—experimental or "vernacular" ecologies. Means and ends are often blurred; either or both can change shape as intentions develop in the course of working out a procedure. Take just one example.

On the derelict Bourn Brook site, just south of the centre of Birmingham, a willow figure has been shaped partly by design (the group had a rough drawing to describe the main structure at the start of the day) and partly through interweavings of people, clay soil, willow wands, and osiers. The rationale for the living sculpture is, like the figure, both determined and undetermined. Determined in the sense that people are here for something in order to affect something. They are determined in that they want to be involved in doing something to express their anger that despite all their efforts it seems as though development will go ahead on this site with little or no attempt to secure its ecological potential. They are local residents, activists from the Birmingham and Black Country Wildlife Trust, people who grew up near the site. They are all trespassing, as this is private land, even though they have used it for years as a place to watch wildlife, to walk in and through, to climb trees, to look upon from home and the nearby allotment. But no one here is quite sure what the action will precipitate, what shape will emerge from the work of hands and hearts. So like the willow figure, the intentions take shape in action. It/they are formed in and through actions that are beyond the design of any one component. To be sure, the structure will not withstand the bulldozers that are set to move onto the site any time now. In this sense you could say that the day's work is an act of senseless beauty (MacKay 1996).

But there is more to ecological attachment than vain gestures. Attachments are forged in action—in this action and in a host of other activities that take place on site (the tree planting, the hedge construction, the tending of allotments, the walks, the fires …). It is almost a cliché to suggest that attachments are forged between people and growing plants as those people take responsibility and care for the seedlings and saplings they have had a hand in cultivating. Perhaps a less telluric way of expressing this is to say that these shared embodiments of people and things heighten awareness, or form a "biopolitical domain" (Thrift 2000). This is far from being an ahistorical form of attachment, with certain privileged kinaesthetic spaces and practices having long and complex evolutions that enfold all manner of nature–culture investments (walking for its own sake being a good example). Suffice to say that biologies and politics both have histories, histories that are entangled with one another. Biopolitical domains are therefore the products of multiple entanglements. So while it is surely accurate to say that human bodies are set up or configured in the world in particular fashions, the fashioning of those worlds can amplify — or otherwise — these configurations. The attachments are therefore part of a mutual pushing that can produce a *"feeling* life (in the doubled sense of both a grasp of life, and emotional attunement to it)" (ibid., 46).

Let us underline again that talk of push and attachment should not be read as a return to an unproblematic sense of present presences. Nor should it be read as a romanticism of local becomings. This is a productive relationship whereby presence and absence can form only part of the story. The feelings and passions that are co-produced in these and other activities are tolerant of be-comings and goings, of likely presences, and are experimental in that no one is quite sure of what will come of the attachment. They are as magical as they are material (and as material as they are magical). Indeed, these and other engagements are perhaps "best understood as a form of magic dependent upon new musics of stillness and silence able to be discovered in a *world of movement*" (ibid., 49; emphasis added). In this sense, the willow figure is neither passive nor strictly active; rather, it is the figuration of something that moves in another voice (Harrison, Pile, and Thrift 2004). Indeed, it embodies the passions of those who have come to value this site and all its connections and movements and who refuse to see it or practise it as dead space awaiting urban renewal.

We have argued that cities are inhabited with and against the grain of urban design, that inhabitants are not static beings but entangled in complex processes of becoming, and that attempts to engage with urban heterogeneity require realignments of people and things in ways that are responsive to uncertainties, indeterminacies, materialities and passions, charm and magic. We have suggested that an ecology that is founded on a straightforward notion of species presence, or what Latour (2004) might call a metaphysics of nature, ill serves the aim of making living cities more livable for people and a multiplicity of others. It is these political implications that we now want to amplify.

Toward a Politics of Conviviality

The notion of living cities fleshes out a sense of ecological co-fabrication in which the life patterns and rhythms of people and other city dwellers are entangled with and against the grain of expert designs and blueprints. This conceptual shift from built environments (as they are termed in conventional town and country planning) to living cities is allied to a realignment of the politics of nature such that cities are appreciated as "ecological disturbance regimes rather than ecological sacrifice zones" (Wolch 1998) in which people are no longer considered inimical to nature, nor natures antithetical to cities. Living with and against the grain of design, and becoming and learning to live among and as others, mark out some of the contours of what we are calling a living city. These challenges to urban analysis call for a re-engaged politics, one that refuses the old settlements between society and nature, between humans and the rest, between matter and mattering: a political re-engagement that we style here as a politics of conviviality that is serious about the heterogeneous company and messy business of *living together*. This politics is gathering momentum through an alliance of urban wildlife and conservation organizations that have gained leverage in the policy process in consequence of the statutory requirement on all Local Councils to produce Biodiversity Action Plans (Harrison and Davies 2002). Just as important, however, it also signals a shift in the politics of knowledge in which expert designs on urban space (including those of conservation science) are more liable to be contested and resisted by city inhabitants (of all kinds) whose ecological vernaculars have been learned and honed through their everyday practices of making themselves at home in the city. We have already suggested that the matters of controversy, the likely presences and the absent presences or urban ecologies, index a fraught political ecology.

To this end we are working toward a conceptual and political style of research that is avowedly and unavoidably a form of intervention in the world, one that opens up rather than pins down the possibilities of city living in play and in prospect. This style, or what we are calling a "politics of conviviality," derives from and informs a triangulation of several theoretical impulses that supplement one another in productive ways and that are all variously in evidence in the heterogeneous passageways and cohabitations explored above in Birmingham. See Figure 4.1.

The first of these impulses concerns the recent reworkings in political theory of the register of "minotarian" politics associated with Deleuze's distinction between micro and macro political levels of analysis: levels that mark significant differences in the kind, rather than the scale, of social attachment and collective action (Deleuze 1995; Deleuze and Parnet 1987). Picking up on his insistence that the dynamics of capitalist culture are uneven all the way down and always already in process, these reworkings point to the proliferation of concrete practices and spaces in which the politics of resistance and difference inhere. This is a politics of association rather than of structure, one in which other possibilities, with and against the grain of any design (or diagram), are always being generated. Notable here is the work of Nikolas Rose (Rose 2000) and James Tully (Tully 1999), who have elaborated a notion of *political agonism* in which they define "citizenship as a capacity to act in relation" (Osborne and Rose 1999, 758), a capacity that is not produced or determined by any one social identity or political alignment but in the multiplicity of relations through which civic associations and attachments are woven. As Rose puts it: "Political agonism is not a traditional politics of the party, the programme, the strategy for the organised transformation of society or the claim to be able to implement...better government. Rather, these minor practices of citizen formation are linked to a politics of the cramped spaces of action on the here and now, of attempts to reshape

FIGURE 4.1
Toward a Politics of Conviviality

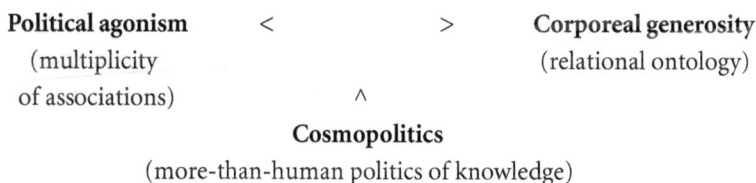

Political agonism	<	>	Corporeal generosity
(multiplicity			(relational ontology)
of associations)		^	

Cosmopolitics
(more-than-human politics of knowledge)

what is possible in specific spaces of immediate action, which may connect up and destabilise larger circuits of power" (2000, 100). In this vein,
the politics of conviviality implicated in our analytical shift toward a "living city" demands that attention be paid to the diversity of ecological
attachments and heterogeneous associations through which the politics
of urban nature is fabricated, as an alternative to reading the political
ecology of the city off a priori or abstract social divisions.

But of what does "acting in relation" consist? How does a politics of
association take, change, and lose shape? The second theoretical impulse
that is important to our project is the one associated with the efforts of
feminist philosophers to elaborate ethical frames that refuse the disembodied and rationalist ontological precepts of liberal individualism
(Whatmore 2002). Notable here is the work of Rosalyn Diprose, who
draws on Merleau-Ponty (among others) to posit an ethics of be-coming
in the world that is thoroughly relational and corporeal, theorizing an
openness to others through which the self is constituted in association as
a necessary and unavoidable condition of social living—what she calls
corporeal generosity. This condition of being given in/to others furnishes
the kind of relational ontology that our politics of conviviality demands
such that "civility" is reconfigured in terms that rely less on excluded
social kinds being "made present" in the civic realm than on the practical
intercorporeality of civic association in which particular kinds or individual entities thrive in combination with others whose capacities and
powers enhance their own. In these terms, our politics of conviviality
share her insistence that the political is ontological (Diprose 2002, 173),
placing onus on the dual role that bodies play in the formation and transformation of civic association and attachment in both the constitution of
different kinds of subjectivities and the situating of these subjectivities in
relation to the bodies of others that we live through. Instead of placing
the political in the realm of conscious judgment and knowledge, here
politics is extended to the "hinterworld of affectivity"—where intercorporeality exceeds the consciousness of "I think" and the "said" of language (ibid., 175).

Finally, the third theoretical impulse that threads through the politics
of conviviality advanced here challenges the residual humanism that is
differently inflected in both the foregoing contributions. It draws on the
work of those philosophers and sociologists of science, notably Isabelle
Stengers (1997) and Bruno Latour (2004), who are seeking to recast the

terms of engagement among science, philosophy, and politics, an endeavour that Stengers calls *cosmopolitics* (Hinchliffe et al. 2005). At least two aspects of this recasting are important to our project. The first relates to the emphasis it places on knowledge as a co-fabrication in which all those (humans and non-humans) enjoined in it can, and do, affect one another in the knowledge event or practice. This involves, as our engagement with black redstarts, the Willow man, and others has illustrated, "the management, diplomacy, combination, and negotiation of human and nonhuman agencies" (Latour 1999, 290). Second, the shift it seeks to make is from a problematic that presumes a gulf between science and politics even as it sets about bridging it, to one that takes their entanglement as given and that redirects attention to the democratization of expertise (Whatmore 2003). Here it is a question of inventing apparatuses such that the "citizens of whom scientific experts speak can...pose questions to which their interest makes them sensible, to demand explanations, to posit conditions, to suggest modalities, in short, to participate in the invention" (Stengers 2000, 160). As Paulson (2001, 112) summarizes: "The crucial point is to learn how new types of encounter (and conviviality) with nonhumans, which emerge in the practice of the sciences over the course of their history, can give rise to new modes of relation with humans, ie to new political practices."

If, as these moves suggest, we accept that research as a knowledge-production process is always, and unavoidably, an *intervention in the world*, then the politics of knowledge at stake here extends to the activity and credibility of social scientists too, demanding that we invest more of our energies in intervening in the terms under which city residents and other urban constituencies are invited, and enabled, to engage in the policy-making process. This obliges social scientists to ally their research efforts and skills in order to experiment with others (humans and non-humans) in making new political configurations possible, in bringing new ecological associations and knowledge practices into being. Such alliances are hard work, take time, and often have to work through suspicions about the intentions and relevance of social-science practitioners. But we think that such interventions in a politics of conviviality are vital to producing politically engaged but "acritical" social science. To be clear, acritical does not mean irrelevant, or even unable to make judgments. Rather, what we want to work toward is a style of research practice that is less hasty in terms of condemning, judging, or looking for the real motivations of people and others. Those motivations, just like the blueprints of organisms, are unfinished expressions that are affected by the actions they partially initiate. To find, work at, and contribute toward events and ethologies that shift the game slightly, that turn presence into likely presence, nature into natures, has been and is our aim.

Conclusion

Our aims in writing this chapter have been to demonstrate a little of the wealth of political ecological practice in cities, and to offer first glimpses of how this practice might be engaged by social scientists. We are suggesting here that the heterogeneity of living cities is different in kind to the cartographies of modern cities and urban theory. For three related reasons. First, cities are inhabited with and against the design of cities. We demonstrated this through a number of examples, all of which — it should be said — conform to particular cartographies of the living and the valued in contemporary ecology (especially in terms of macrofauna and the rare). We have mentioned in passing the multiplicity of non-humans that inhabit cities, but there is of course more to be said here (see Mabey 1999 on this, and Dion and Rockman 1996 for a quirky transect through the urban jungle). Second, cities are not simply inhabited but cohabited, in ways that are multiple and entangled and that disrupt established ethologies and ecologies. Animals, plants, microbes, and the multiple relations within and among these temporary stabilizations, become urban, often in ways that are surprising. Third, engaging with these inhabitants and becomings requires political and scientific experiments, relaxing the coordinates of presence and absence that are so dominant in scientific and legal conservation theory and so ungainly in practice. In the final part of the chapter we have suggested three overlapping but by no means mutually consistent strands of theoretical practice that can offer resources for engaging living cities. We have drawn on understandings of political agonism, corporeal generosity, and cosmopolitics in order to point toward a different kind of politics that may well be more suited to the ecologies we are tracing. In interweaving empirical work with humans and non-humans, and rereadings of political and social theory, our aim has been to highlight the difference of, and to make a difference to, living cities.

Acknowledgments
The authors acknowledge the support of the Economic and Social Research Council, which funded the research project from which this paper derives (Award no. RA00239283).

Works Cited

Ansell Pearson, K. 1999. *Germinal Life: The Difference and Repetition of Deleuze.* London: Routledge.

Babtie, G. 2001. "New Hospital Proposal: Water Vole Survey." Glasgow: Babtie Multidisciplinary Consultants.

Barker, G. ed. 2000. *Ecological Recombination in Urban Areas.* Peterbrough: Urban Forum/English Nature.

Davis, M. 2002. *Dead Cities.* New York: New Press.

Deleuze, G. 1995. *Negotiations.* New York: Columbia University Press.

Deleuze, G., and C. Parnet. 1987. *Dialogues.* New York: Columbia University Press.

Dion, M., and A. Rockman, eds. 1996. *Concrete Jungle: A Pop Media Investigation of Death and Survival in Urban Ecosystems.* New York: Juno.

Diprose, R. 2002. *Corporeal Generosity: On Giving with Nietzsche, Merleau-Ponty, and Levinas.* New York: SUNY Press.

Harrison, C., and G. Davies. 2002. "Conserving Biodiversity That Matters: Practitioners' Perspectives on Brownfield Development and Urban Nature Conservation in London." *Journal of Environmental Management* 65: 95–108.

Harrison, S., S. Pile, and N. Thrift. 2004. "Developing Patterns: Ethologies and the Intense Entanglements of Process." In *Patterned Ground: Entanglements of Nature and Culture*, ed. S. Harrison, S. Pile, and N. Thrift. London: Reaktion.

Hinchliffe, S., M. Kearnes, M. Degen, and S. Whatmore. 2005. "Urban Wild Things: A Cosmopolitical Experiment." *Environment and Planning D: Society and Space* 23, no. 5: 643–58.

Hough, M. 1995. *City Form and Natural Process.* London: Routledge.

Latour, B. 2004. *Politics of Nature: How to Bring the Sciences into Democracy,* Cambridge, MA: Harvard University Press.

———. 1999. *Pandora's Hope: Essays on the Reality of Science Studies.* Cambridge, MA: Harvard University Press.

———. 1993. *We Have Never Been Modern.* Hemel Hempstead: Harvester Wheatsheaf.

Mabey, R. 1999. *The Unofficial Countryside.* London: Pimlico.

MacKay, G. 1996. *Senseless Acts of Beauty: Cultures of Resistance.* London: Verso.

Matthews, A. 2002. *Wild Nights: The Nature of New York City.* London: Flamingo.

Osborne, T., and N. Rose. 1999. "Governing Cities: Notes on the Spatialisation of Virtue." *Environment and Planning D: Society and Space* 17: 737–60.

Paulson, W. 2001. "For a Cosmopolitical Philology: Lessons from Science Studies." *SubStance #96* 30, no. 3: 101–19.

Rose, N. 2000. "Governing Cities, Governing Citizens." In *Democracy, Citizenship, and the Global City*, ed. E. Isin. London: Routledge.

Serres, M. 1995. *The Natural Contract.* Ann Arbor: University of Michigan Press.

Stengers, I. 2000. *The Invention of Modern Science.* Minneapolis: University of Minnesota Press.

———. 1997. *Power and Invention: Situating Science.* Minneapolis: University of Minnesota Press.

Thrift, N. 2000. "Still Life in the Nearly Present Time." *Body and Society* 6, nos. 3–4: 34–57.

Tully, J. 1999. "The Agnonic Freedom of Citizens." *Economy and Society* 28, no. 2: 161–82.

Whatmore, S. 2003 "Generating Materials." In *Using Social Theory*, ed. M. Pryke, G. Rose, and S. Whatmore. London: Sage.

———. 2002. *Hybrid Geographies: Natures, Culture, Spaces.* London: Sage.

Whatmore, S., and S. Hinchliffe. 2003. "Living Cities: Making Space for Urban Nature." *Soundings* 22: 137–50.

Wheater, P. 1999. *Urban Habitats.* London: Routledge.

Wolch, J. 1998. "Zoopolis." In *Animal Geographies*, ed. J. Wolch and J. Emel. *Animal Geographies.* London: Verso.

Wyatt, N., ed. 2003. *A Provisional Atlas of the Mammals of Birmingham and the Black Country.* Birmingham: Wildlife Trust for Birmingham and the Black Country.

Part Two

Experiencing Technonatural Cultures

Chapter Five

Boundaries and Border Wars: DES, Technology, and Environmental Justice
Julie Sze

According to a recent study, newborn babies in the United States have absorbed an average of two hundred industrial chemicals and pollutants through the mother's womb. Their umbilical cord blood contains pesticides, consumer-product ingredients (including Teflon), stain and oil repellants from fast-food packaging, clothes, and textiles, and wastes from coal, gasoline, and garbage (EWG 2005). Even in a highly polluted society, the extent to which newborns come into the world with a "human body burden" comes as disturbing revelation for environmental activists and health researchers. That newborns enter the world marked by pollution highlights the contradiction between the idealized notion that babies are innocent and pure and the reality that they are born in and of the toxic soup that comprises the post–Second World War landscape of pesticides, chemicals, and plastics. In other words, polluted babies are troubling technonatural creatures because they collapse the boundaries of the bodily and the natural with the technological, the man-made, and the synthetic.

Arguably, the first well-known example of the problems of synthetic intervention during pregnancy and childbirth arose with the widespread use of diethylstilbestrol (DES), a man-made estrogen discovered in 1938 in Britain by Charles Dodds. It was "a novel and a daring product" that produced the same feminizing effects as estrogens derived from plants and animals, but that was three times more powerful (Seaman 2003, 36). DES was a man-made version of a "natural" hormone, and because it was derived "from nature," it was presumed to be beneficial. In reality, DES was toxic for the main populations to which it was given: women and livestock animals. Between four and six million women in the United States from 1948 to 1971 were prescribed DES to prevent miscarriages (Apfel and Fisher 1984, 1). It was promoted by pharmaceutical companies and prescribed by medical doctors as a "miracle drug" for treating menopause and drying up breast milk in non-nursing mothers. It was also used as a "morning after" contraceptive, besides being given to girls growing "too tall" and to male-to-female transsexuals before sex-change operations (Seaman and Seaman 1977). Moreover, it was standard practice for farmers to give DES to fatten chickens, cows, and other livestock and to render their meat more succulent; it was also a way to chemically castrate male livestock (Marcus 1994, 12–13). Later, mothers given DES during pregnancy were suspected as having a higher risk of breast cancer, and DES-exposed children were shown to face a higher risk of various cancers and genital abnormalities — the first known human occurrence of transplacental carcinogenesis (cancer-causing effects) as a consequence of in utero exposure (DES Action 2008).

In this chapter I build on Donna Haraway's articulation of the cyborg as a cybernetic hybrid of organism and machine, to argue for an analysis of DES and a revision of the cyborg concept through the framework of "technologically polluted bodies." DES offers a unique prism through which to understand nature's social construction through technology and the human body, as opposed to the wilderness as the idealized site of "nature" in environmental studies. As was outlined in the introduction to this book, the nature of nature is increasingly up for grabs; I would note here that that contestation has a past. Environmental historians have questioned the centrality of the idea of "nature" as pristine green space absent of people, as well as the racial and gendered implications of that construction (Cronon 1996; Merchant 1996; Spence 1999). The emerging social movement and academic field of environmental justice offers another challenge to the "nature of nature" as a category unmarked by race or class. At the core of the term "environmental justice" is a redefinition of the "environment" to mean not only "wild" places but also the environment of human bodies, especially in racialized communities, in cities, and through labour (exemplified by the movement's slogan that the environment is where people "live, work, play, and pray"). The definition and redefinition of "nature" and "environment" are thus cultural tasks, in which narratives and stories form a central part (Stein 2004; Adamson, Evans, and Stein 2002).

Using DES as the case study, I advance an understanding of technologically polluted bodies, not as Haraway's hybrids of organism with machine, but rather as hybrids of bodies — animal and human, and particularly *female* — with non-machine–based forms of technological intervention such as the pharmaceutical, petrochemical, and livestock industries and the products they create and normalize through their production processes. DES provides a rich example of how human and animal bodies interact with a variety of technological and environmental systems. To understand DES bodies as technologically polluted is to argue, as Haraway suggests, against the purity and integrity of social, natural, and bodily categories and in favour of what she calls "boundary breakdowns."

As a case study in polluted women and livestock (animal bodies), DES illustrates changes in the human relationship to nature and what these changing relationships might mean for the possibility of justice and ethics in a hyperpolluted, highly technological world of corporate concentration. If we are what we eat in food and medicine (DES in animal bodies consumed as food for humans, and given to women as medicine), then what we eat alters our body in a feedback loop that calls into question any idea of the body or nature that is pure or unadulterated. DES bodies illuminate how we are already always hybrid and that there is no nature or body that is not shaped by culture, technology, or medicine, no purity on which we can stand in order to define concepts of nature, race, gender, or humanity itself. The ubiquity of border crossings, hybridity, and cyborgian alterations makes a politics or philosophy of justice and/in technology even more urgent.

Female Bodies and Scientific Bodies of Evidence

DES, a man-made non-steroidal estrogen, was synthesized in 1938 by the chemist–physician Edward Charles Dodds in London (Fenichell and Charfoos 1981). Derived from coal tar, it differs from steroidal estrogen in that it lacks the four interlocking carbon rings that characterize natural steroid hormones and their derivatives. As Cynthia Orenberg (1981) explains, DES lacks the chemically distinct four-ring structure. DES is a "ringer" and "*fundamentally different*" in its chemical structure. Though it behaves empirically "like estrogen," it isn't: it is three times more powerful, and it is not destroyed or affected by gastric secretions such as natural estrogen. For Orenberg, the central paradox is that DES was "accepted as the real thing" by physicians who "never thought that they might be tampering dangerously with nature" (1981, 11). Dodds "resolved to diverge radically from nature in order to mimic it" (Fenichell and Charfoos 1981, 17). He was also an inherently "conservative" physician, who rejected the idea of DES as a "miracle" cure. He expressed "humble" respect and awe for the female reproductive body. He noted that male workers in his laboratory who handled DES grew female breasts and became impotent. Eventually, DES was produced by women to avoid this problem (ibid., 17). In an 1938 report in *Nature,* he expressed his suspicion of another problem: that being highly estrogenic, DES was also carcinogenic (ibid., 16). By 1939, 40 articles had been published demonstrating the carcinogenic effects of synthetic and natural estrogens in animals; several of these focused specifically on DES. By 1941, 257 papers had demonstrated the value of DES for treating menopausal symptoms and for other uses. These studies lacked adequate controls and were not "double-blind." Despite inadequate research data, in 1952 the FDA by administrative fiat declared DES to be safe and no longer requiring annual approval (Dutton 1988).

Though DES was first synthesized in London, its widespread popularity can be considered an American phenomenon stemming from the crucial role played by American doctors associated with elite medical institutions and pharmaceutical companies. Dodds had conducted his research within the British system and thus was prohibited from taking out a patent on DES, and he endorsed the idea that his invention ought to stay within the public domain so that it could be used for the "greater good of humanity," especially since he had created DES in a wartime race with the Nazis and their eugenics program (Seaman 2003). No restraints held back the pharmaceutical companies, which distributed samples widely and encouraged research on DES's "miraculous effects." The widespread use of DES during pregnancy began in 1947, a result of the efforts of a husband-and-wife research team at Harvard Medical School, George and Olive Smith. Olive Smith's 1948 groundbreaking paper and their 1949 follow-up in the *American Journal of Obstetrics and Gynecology* encouraged DES's use to prevent miscarriages (Fenichell and Charfoos 1981). According to their theory, elevated estrogen levels during pregnancy stimulate progesterone, which is essential for the uterus to receive and sustain the egg. Inadequate levels of either led to complications or failure of the pregnancy. At a 1949 medical meeting the Smiths declared that DES benefited *all* first-time pregnancies. In their words, DES seemed to "render normal gestation 'more normal'" (Dutton 1988, 54). DES was considered benign because it was making a "natural," "biological," and "normal" process more effective (Apfel and Fisher 1984). That DES was relentlessly promoted exemplifies a utopian belief that technologies could harness and "improve" on nature itself. While this belief system was not unique to the United States, certain factors made it particularly dangerous there—specifically, the pharmaceutical industry's marketing power and its tactics aimed at weakening the Food and Drug Administration (FDA).

The Smiths' theory was refuted by a 1953 University of Chicago study by William Dieckmann, which definitively found that DES was *ineffective* at preventing miscarriages. (Indeed, later reanalysis of the data found that DES was actually linked to *increased* miscarriages, premature births, and higher infant mortality; Orenberg 1981, 4.) Yet doctors continued to widely prescribe DES in normal pregnancies as if it was a "vitamin," as "a little extra insurance." They continued because of the highly aggressive sales tactics of pharmaceutical representatives: DES by then was being widely advertised in medical journals and the popular press. Influenced by these ads, and desperate to avoid miscarriages, women were demanding DES. One 1957 ad featured a happy baby and read: "[DES is] recommended in ALL pregnancies... desPLEX tablets also contain vitamin C and certain members of the vitamin B complex to aid detoxification in pregnancy and the effectuation of estrogen" (Apfel and Fisher 1984, 26). Pregnant women continued to use DES heavily until a 1971 study confirming the link between in utero DES exposure and a rare vaginal cancer— clear-cell adenocarcinoma—was published in the *New England Journal of Medicine*. Only then did the FDA issue an alert about DES use in pregnancy. Even after DES's link with cancer was shown, it was commonly used as a "morning-after pill" on university campuses well into the 1980s, though the FDA never approved it for that use (Dutton 1988). According to the DES Cancer Network, none of the 267 pharmaceutical companies that produced and distributed DES have accepted any responsibility for DES's health effects.

Reading DES and Gender

Existing histories of DES analyze it through the lenses of medical history and the history of regulation in the livestock or pharmaceutical industries, and ignore certain cultural questions. There is, in other words, no parallel to Rachel Carson's incisive critique of DDT and pesticides in postwar American culture as captured in her powerfully influential book *Silent Spring* (1962). Rather, what we have is what science-and-technology scholar Joseph Dumit describes as a set of histories that detail "[DES's] incredibly tragic history within a kind of enlightenment narrative. They state that DES was not studied carefully enough as first, and those studies which showed problems were ignored by the medical community at large. When irrefutable proof of DES's harm was provided in 1981 (the narrative goes) the medical community responded, the public was outraged, and more research was conducted" (Dumit with Sensiper 1998). These accounts treat DES in animals and women along parallel tracks, making its timeline difficult to decipher (DES was banned in chickens in 1958, yet its use in cattle continued until 1979 and its use in women continued through the 1980s). They generally portray DES as a tragic or peculiar historical episode that tells a particular tale about medical knowledge or about the history of government regulation (Apfel and Fisher 1984). For others, DES is "a modern meat production milestone, perhaps the most important single occurrence in the chain of events that culminated in the current methods of production" (Marcus 1994).

The particularly American aspects of DES tell a story about the relationship between production and dissemination — a story that, as Alan Marcus suggests, emerges from a complicated dance between corporate and academic scientists and government regulators. The use of DES in beef arose from a researcher at Iowa State University, its use in chickens from the University of California at Davis (ibid.). The corporate–academic–government nexus relating to DES emerged specifically from an American model of research that flourished in the postwar era. How can the power of the pharmaceutical and agriculture lobbies and the inefficacies of the FDA and the U.S. Department of Agriculture (USDA) as regulatory agencies be related to each other? These kinds of questions cannot be asked or answered using existing analytic frames. And one cultural question is almost never asked: How can we think about the relationship between women and animals *through* DES?

When histories of DES focus on women, they tend to downplay the complexities of gender. Popular and historical accounts of DES's use in women deploy an early feminist framework that emphasizes women's victimization (Apfel and Fisher 1984). Orenberg (1981) describes DES as "her story" as well as that of her daughter, who was exposed in utero. DES is "only one example of the consequences of thinking that modern medicine is infallible, that the physician is sacrosanct, and that the patient (particularly the *women* patient) is an object to be 'done to'" (1981, 28). DES represents prevailing medical practices and the community's attitude toward research on women without their consent, the "chilling number of trial-and-error medical experiments using DES on women" (1981, 28). Personal narratives are gendered insofar as they focus on the perspectives of the women who took DES or on daughters who were exposed to it in utero, whose genitalia were arguably more altered than those of DES sons and whose cancer risk was higher (Davidson 2003). Lastly, the FDA's continued inaction on DES was successfully challenged by the women's health movement, particularly by DES-exposed mothers and their children (Braun 2001). Women's magazines, which had promoted DES earlier, later became effective venues for communicating its risks (Dutton 1998, 75–76).

What these narratives and histories lack is a sustained consideration of how gender and sexual development as constructed categories can illuminate the cultural significance of DES. For example, in 1958, DES use was suspended in poultry after cases of early sexual development in young children in Puerto Rico and Italy were correlated with high chicken consumption (Apfel and Fisher 1984, 15). Also, male farm workers who fed DES to chickens and men who ate large amounts of chicken developed breasts, reported sterility, lost facial hair, and developed high-pitched voices (ibid.). What made DES's use in chickens unpalatable (literally) was the way in which it visibly made men "women" and made children sexually mature. The "unnatural" sexual and bodily developments that DES triggered were visible, embodied, and therefore grotesque. When these categories of gender and sexual development were made manifest, DES was banned. Though the long-term harms of DES in other livestock and in women were as harmful, they were not visible in quite the same way and were not acted on until the political and cultural climate changed and there was a greater body of scientific evidence of DES's harms.

DES also highlights how *female identity* itself was defined medically and socially through hormones. If being female and male can be defined through hormones, then a whole host of female and male "problems" can be solved. So goes the logic. Throughout the twentieth century, female hormones were given to categories of women who were considered insufficiently female, such as menopausal women and lesbians (Seaman

2003, 54; Serlin 2004). The larger epistemological and cultural questions —
Why were these "conditions" considered medical problems in the first
place? Why were they seen as requiring medical intervention? — have
been explored by scholars in queer studies and intersex studies. This
same line of critical interrogation has analyzed the parallel "problem" of
the menopausal woman, since the first popular and regulated use of DES
for its "miraculous" effects on menopausal women. The search for a remedy for menopause has been replete with cultural stigmas about female
identity in the post-reproductive years. One popular book in the 1960s by
Robert Wilson titled *Feminine Forever* stated that "a woman is not 'complete'" unless she takes hormone replacement pills and that she will be
"condemned to witness the death of her own womanhood" (discussed in
Seaman 2003, 55). National ads in the 1950s and 1960s for hormone
replacement therapy were often openly sexist, depicting menopausal
women as "repulsive, witchlike… angry or depressed, menacing" and as
prone to violence once their reserve of estrogen was gone (ibid., 49).

The most obvious way in which gender shaped DES was through gen-
der-normative definitions of a woman's identity through pregnancy and
childbirth. Miscarriage was to be avoided at all costs, at least in part
because women defined their female identity in terms of successful fertility, childbirth, and, ultimately, motherhood. Thus, with the goal of carrying a successful pregnancy and birth, millions of women took a drug
derived from their natural hormones, a drug that ultimately led to their
daughters having a higher cancer risk as well as drug-altered uteruses that
made their own pregnancies difficult and dangerous.

Cultures of DES: Technology, Cyborg and Hybrid Stories

Besides rereading DES through critical frames of American values and
corporate culture, gender construction and sexual identity, we need to
consider it from the perspective of technology studies. Is DES a reproductive technology? (Holmes 1992). One feminist critic has noted that "technological interventions in the womb are extraneous parties (*objects or
people*) that hinder, modify, or enhance female reproduction" (Lublin
1998; italics added). Though most descriptions of reproductive technology do not include DES, it *does* fall under a broad definition of technology in general and reproductive technology specifically as an application
of scientific knowledge to assist in making babies (in this specific case, by
supposedly preventing miscarriages).

Ellul's (1964) definition of "technique" is illuminating in that it reframes DES vis-à-vis technology studies. He writes that though we automatically link technology with machines, "it is a radical error to think of technique and machines as interchangeable" (1964, 7). Rather, "technique does not mean machines, technology or this or that procedure for attaining an end. In our technological society, technique is the totality of methods rationally arrived at and having *absolute efficiency*" (ibid., xxv). Broad definitions of reproductive technology that include attempts to "improve" the pregnancy thus include DES under its umbrella. Expanding the definition of reproductive technology to include DES also draws the pharmaceutical sector into a history that predates in vitro fertilization and other more easily recognized reproductive technologies (ibid., x). DES is neither object nor person, but its function was indeed to "enhance" female reproduction. In the spirit of the Smiths and a generation of doctors treating pregnant women, it was a tool for achieving "better," more natural and normal pregnancy—in Ellul's words, the goal of *absolute efficiency* in pregnancy and childbirth. DES—as technology and technique—complicates definitions of technologies as object or machine-based.

Reconsidering DES vis-à-vis technology studies also situates it within the framework of Haraway's influential cyborg theory. As a reconsideration of the cyborg, DES provides an ideal case study for understanding how to integrate cultural theory with pedagogy and activism. I use Haraway's cyborg theory to suggest how to teach DES in a classroom using literary analyses. My chosen text is Ruth Ozeki's *My Year of Meats* (1998). The novel is narrated by Jane Takagi-Little, a "DES Daughter." Jane is a mixed-race (Asian/White, Japanese/American) aspiring documentary filmmaker who at the beginning of the story is a corporate tool for a Japanese TV show, *My American Wife!*, sponsored by the American meat export lobby to increase its sales in Japan. I focus on this novel because it resurrects the DES story long hidden from popular consciousness. Like Upton Sinclair's *The Jungle*, it is a novel with muckraking intentions. It also raises complex themes of hybridity and of the cyborg that can illuminate larger questions about culture, technology, and the body, and it does so in an accessible text that is easily taught in undergraduate classrooms.

By focusing on the cultural production of the cyborg as a "creature of fiction," and on our collective complicity in the existence of cyborgs, Haraway places the agency, pleasure, and politics of their construction and existence in our own hands (1991, 150). As a creature of "fiction," the cyborg is located where the appropriation of nature and the production of culture meet and mesh, in which "the stakes in the *border war* have been the territories of *production, reproduction, and imagination*" (ibid.). Her focus on these intertwined realms is echoed by Dumit in his call for an investigation of the epistemology of the "facts" of DES (Dumit with Sensiper 1998, 216). A fact is "a word used to describe the situation where (our) culture and nature agree. To call something a fact is to represent a cultural consensus on the nature of nature." Dumit argues that drug-altered bodies — in particular, DES bodies — are cyborg and that more research needs to be done on how and why original cyborgian alteration with drugs extends to further alterations with technologies (ibid.). If DES bodies are cyborgian, what are the politics of that identity? Haraway suggests that the cyborg myth is about "transgressed boundaries, potent fusions and dangerous possibilities which progressive people might explore as one part of needed political work" (1991, 154). Three boundary breakdowns are central to understanding cyborgs and the boundaries they transgress: the breaching of the boundary between human and animal, between animal–human (organism) and machines, and between the physical and the non-physical. In particular, "the dualisms of self/other, mind/body, culture/nature, male/female, civilized/primitive, reality/ appearance, whole/part, agent/resource, maker/made, active/passive, right/wrong, truth/illusion, total/partial, god/man" are challenged (ibid., 151). Transgression enables freedom from epistemological and historical constraints and dominations because cyborgs complicate long-standing dualisms that have functioned to dominate women, people of colour, nature, workers, and animals. For Haraway, "cyborg politics is the struggle for language and the struggle against perfect communication, against the one code that translates all meaning perfectly... That is why *cyborg politics insists on noise and advocates pollution, rejoicing in the illegitimate fusions of animal and machine*" (ibid., 177; italics added).

In thinking about how Haraway's cyborg analysis can be used to understand DES, several questions emerge that confirm and possibly trouble her analysis. Are "illegitimate fusions" of animals and machines necessarily something to "rejoice"? Is the pleasure inherent in the transgression, in the border crossing, and in the collapsing of categories? Why do cyborg politics necessarily "advocate pollution?" What does this "pollution advocacy" mean vis-à-vis the *actual* case of DES? Do technologically polluted bodies merely represent a naive wish to return to a pre-cyborgian, pre-hybrid state of the unpolluted body? What are the "dangerous possibilities" that cyborgian transgressions, boundary breakdowns, and rejections of dualisms represent, if we are to take up Haraway's call as "progressive people" exploring "political work"? These are questions I now turn to through a close analysis of DES in *My Year of Meats*.

DES Narratives: Gender, Race, and Hybridity in *My Year of Meats*

As Haraway suggests, the "border war" between nature/culture and environment/technology is contested on the terrains of production, reproduction, and imagination. Thus, production/reproduction/imagination is a useful frame for analyzing cyborgs and, by extension, DES as a cultural narrative. From photography collections to documentaries and plays, DES has been a topic of cultural production (Braun 2001). This focus is not surprising, because the personal stories these cultural productions reveal serve as an important counternarrative to the overwhelming statistical and medical tone of dominant DES narratives. What sets *My Year of Meats* apart from most DES histories is its dual focus on women *and* animals. Takagi-Little is a "DES daughter" who works for *My American Wife!* which highlights a variety of American women, their families, and their meat dishes from particular regional and ethnic subcultures in the United States. The aim of the show is to increase American meat sales in Japan by teaching Japanese women how to cook unfamiliar kinds of meat.

As a racially mixed DES daughter, Jane embodies Haraway's critique of static categories of identity, specifically of racial and national identities. The author ties Jane's mixed-race status to two key terms in the novel: hybridity and sterility. As a person of mixed race and binational heritage, Jane acts as a cultural broker between two nations and cultures. As she explains, "being racially 'half' — neither here nor there — I was uniquely suited to the niche I was to occupy in the television industry... Although my heart was set on being a documentarian, it seems that I was more useful as a go-between, a cultural pimp, selling off the vast illusion of America" (9). While Jane's mixed-race status gives her authority as a cultural broker between nations and races, Ozeki also ties her status to non-human (specifically plant) hybrids. Hybrids are clear rejections of "nature." But this rejection is complex for Jane. Neither unproblematic nor idealized, hybrids stand as both a *warning signal* and an *opportunity* to escape a cultural past obsessed with notions of purity. Thus, race and culture are likened to native species that are increasingly moving into non-native locales. For Jane, the cautionary tale of this crossing is to be found in the story of kudzu, an invasive plant species that represents the dangers of careless botanic transplantation and "biological invasions." The discourse of biological invasion has been tied to social ideas and cultural anxieties about place, nature, and culture, particularly in the South (Buh 2004). Kudzu is a Japanese plant, touted as a "miracle plant" in the United States, that was brought to the American South to "rescue" its depleted soils. But it soon overran more than 500,000 acres in the South, owing to its "predaceous and opportunistic" and fast-growing nature (it often grows a foot a day). It echoes DES itself, and as such it represents the unforeseen consequences when miracles (be they technological or botanical) go awry. Jane describes how kudzu has been used as a disparaging metaphor by American nativists for the economic "invasion" of the Japanese in the South (77). At the same time, the movement of plants and people represents an outcome she welcomes: "All over the world, native species are migrating, if not disappearing, and in the next millennium, the idea of an indigenous person or plant or culture will just seem quaint. Being half, I am evidence that race, too, will become a relic... Some days, when I'm feeling grand, I feel brand-new—like a prototype... Now, oddly, I straddle this blessed, ever-shrinking world" (15). Kudzu is an ambiguous metaphor for both Jane and the author, representing both freedom and danger in the flows and movements of peoples, plants, and cultures around and within the world.

Jane's racial hybridity is also linked to her fertility, both actual and perceived and—like the discussions of human and plant hybrids in the novel—complex and ambivalent. On the one hand, Joichi Ueno, the Japanese advertising executive in charge of the Beef-Ex account and the show, says in a moment of drunken flirtation: "You, Takagi, are a good example of hybrid vigor... We Japanese get weak genes through many centuries' process of straight breeding. Like old-fashioned cows. Make weak stock. But you are good and strong and modern girl from crossbreeding" (43). Part of her perceived "strength" comes from her height—she is taller then most Japanese women, more direct, and socially nonconformist. "Hybrid vigour" is a well-known term among breeders and farmers that applies to the "exceptional sturdiness" of first-generation crossbreeds in both animals and plants. Thus, the author uses Jane's racial hybridity to stand for agricultural and non-human "natural" ideas of fertility and strength. Jane's "hybrid vigour" is contrasted with the racial purity of her narrative double, Ueno's wife, Akiko. Their marriage is plagued with problems, in part because of Akiko's infertility and his "stoney rage." The irony is that Jane has more difficulties getting pregnant and carrying a pregnancy to term because of the structural changes in her uterus as a result of her DES exposure, which she discovers in the course of the narrative. Her infertility caused the fractures that led to her first divorce and is linked to her mixed-race status, because "like many hybrids...I was destined to be nonreproductive" (152), referring to the tendency of certain hybrids (such as mules) to be sterile. Jane's infertility is doubly ironic: her mother had taken DES for the sake of being successfully fertile; at the same time, it is supposed that Jane's strength is a result of her racial hybridity. Jane expresses the freedoms that come with hybridity—freedoms that Haraway suggests are enjoyed by cyborgs. But alongside the geographic, biological, racial, and national freedoms come dangers, ambivalence, and complexity. With hybridity, perceptions of hyperfertility and strength work alongside themes of infertility and sterility.

Technologically Polluted Bodies: We Are What We (M)eat

While Jane embodies Haraway's notion of a racial and technological hybrid and cyborg, a close reading of *My Year of Meats* suggests that critiques of fixed categories of identity that cyborgs implicitly embody need not lead to a celebration of technologically polluted bodies. Haraway's cyborg theory provides a crucial and important critique of fixed identities and in doing so may make it difficult to critique pollution. Ozeki's novel explicitly challenges this stance. The key to this challenge is how Jane is further shaped and altered through what she eats—specifically, *meat*—and what she learns about the industry, which uses hormones like DES to "improve" their products. In the novel, Ozeki connects women and livestock in a complicated stance that simultaneously embraces hybridity and rejects pollution.

As the novel opens, Jane is the coordinator of the production team for *My American Wife!* and in that capacity interprets the directive from the Tokyo office on behalf of the show's sponsor (Beef-Ex) to remember that "Beef Is Best." As she writes in a memo describing the show: "Meat is the Message. Each weekly half-hour episode of *My American Wife!* must culminate in the celebration of a featured meat, climaxing in its glorious consumption. It's the meat (not the Mrs.) who's the star of our show! Of course, the "Wife of the Week" is important too. She must be attractive, appetizing, and all-American. She is the Meat Made Manifest: ample, robust, yet never tough or hard to digest" (8).

Meat, in other words, is how ultimately one can make sense of DES's rise in the United States, both as *a symbol* connecting women and animals and as a *technological process* to *control nature* and *maximize efficiency* through technology. Anthropologists have long analyzed meat's cultural and symbolic importance, arguing that it represents the domination of humans over nature and non-humans (Fiddes 1991). Thus it comes as no surprise that meat is so central to the novel, given that it acts a rich signpost of cultural and moral values. As Jane writes when describing the American Wives on the show: "Through her, Japanese housewives will feel the hearty sense of warmth, of comfort, of hearth and home — the traditional family values symbolized by red meat in rural America" (8).

Yet at the same time, the darker side of the "values" of meat is represented by DES. Thus women and animals are linked in DES, not accidentally or incidentally, but through an American technological and medical culture that sees the improvement of nature through technology and increased efficiency as central to the larger cultural project of improvement and progress. Furthermore, the links between reproduction and production and between animals and women perform and create *new kinds* of technological violence in the bodies they inhabit. These links between animals and women are not surprising, because, as Haraway notes, one of the key boundary breakdowns characterizing the cyborg is the human/animal breakdown. Furthermore, the boundary breakdown between *women* and *animals* has been a central feature of the history of hormone development, in that animals have been central to the development of the birth control pill and of reproductive technologies such as IVF and hormone-replacement therapy, besides being the sources of popular estrogen therapies. Pioneers of IVF, such as Jacques Testart in France and Alan Trounson in Australia, began their careers as animal biologists and often sexualized their control over female fertility (Raymond 1994). DES was first used as a treatment for menopause, but it was soon supplanted by Premarin, the most commonly prescribed drug for treating menopause, introduced in 1942. Premarin is derived from a pregnant mare's urine during the third through tenth months of the equine gestational period (hence the name *pregnant mare urine*). An estimated 35,000 mares stand in barns throughout Canada and parts of the American Midwest for about six months out of every year with urine-collection devices strapped onto them. Most of their foals are sold for slaughter.

In making animals meat, and female reproduction increasingly efficient, DES stands as a symbol not only for pollution in the processes of meat and baby making, but also for changing systems of production more generally. Meat functions as a larger symbol of forms of racialized American violence and for violence against animals in meat production. Jane describes the well-publicized murder of a Japanese exchange student, Yoshihiro Hattori, who was shot to death by Rodney Dwayne Peairs when he rang on a doorbell to ask for directions to a Halloween party. Jane notes that Peairs worked as a meat packer in Louisiana: "Hattori was killed because Peairs had a gun, and because Hattori looked different. Peairs had a gun because... we fancy that ours is still a frontier culture, where our homes must be defended by deadly force from people who look different. And while I'm not saying that Peairs pulled the trigger because he was a butcher, his occupation didn't surprise me. Guns, race, meat, and Manifest Destiny all collided in a single explosion of violent, dehumanized activity" (89).

Though numerous scholars have linked notions of the frontier, Manifest Destiny, and American imperialism to racialization, Jane adds meat and systems of its production and consumption to this cultural history. This has contemporary "real world" implications for how people (especially disenfranchised people, such as people of colour and immigrants) live, work, and consume meat. According to a Human Rights Watch report (2005), meat-packing is the most dangerous industry in the United States, with few worker protections. That industry is also increasingly populated by immigrant workers. In this sense, the novel expands on Roger Horowitz's (2003) examination of race, gender, ethnicity, and technology in the American meat-packing industry. Horowitz asks, but does not answer: Is there a role for the household or the consumer in this story? My Year of Meats offers precisely that view of race, gender, and ethnicity in the household and consumption aspects of meat production in the United States (2003).

Individual violence, the example at hand being Hattori's murder, serve as microcosms of larger cultural forms of systemic and technological violence enacted through meat production. The health effects of hormones, drugs, and chemicals are countless, and the cruel conditions suffered by animals (overcrowding, the cutting off of chickens' beaks) are well documented. The conditions in which the animals live in the livestock industry are a logical outcome of the story that began with DES—that is, with the changing technologies of food production and consumption. As Jane describes: "DES changed the face of meat in America. Using DES and other drugs, like antibiotics, farmers could process animals on an assembly line like cars or computer chips. Open-field grazing for cattle became inefficient and soon gave way to confinement feedlot operations or factory farms, where thousands upon thousands of penned cattle could be fattened at troughs. This was an economy of scale. It was happening everywhere, the wave of the future, the marriage of science and big business" (125). Meat, in other words, became increasingly mechanized, animals became things, economies of scale grew, and older forms of food production and consumption were abandoned.

Thus the key cultural issue that *My Year of Meats* considers is not whether eating meat is natural/right/ethical. Rather, the issues are these: How has meat been made *differently* technological? What food and social systems have developed over the past fifty years that are significantly different in scale and scope from older systems of production and consumption? And what do race and gender have to do with these changes? (Ecofeminists like Carole Adams [2000] argue that these racialized and gendered politics extend back to colonial contact.) What has emerged is a production system that has completely mechanized its product—in other words, that has taken animals away from nature and into technology. As Horowitz (2006, 151–52) suggests: "Convenience [in meat production] ...rested on ever greater intervention into *nature*...Implicit in the very notion of convenience was using *technology* to help mankind claim victory over the organic subduing animals and their parts to the imperative of the human race. Altering animal biology and growth patterns, tinkering with forms of processed meat, adding chemicals to feeds, creating more automated production methods—all were elements of relentless efforts to turn *nature's bounty into products that fit with the modern lifestyles of our civilization*" (italics added). DES is simply one especially salient (salacious?) example of this tension between nature and technology, one that reflects cultural desires of control and domination that shape the particular contours of production.

Yet another logical outcome of these changes in production and its utter moral bankruptcy is represented in *My Year of Meats* by Gale Dunn, the son-in-law of Bunny Dunn, a featured "Mrs." on *My American Wife!* and wife of Colorado rancher-patriarch John Dunn. To Jane, Gale describes the cattle feed as "recycled": "We even got by-products from the slaughterhouse — recycling cattle right back into cattle. Instant protein... The formulate feed we use is real expensive, and the cattle shit out about two-thirds before they can even digest it. Now there's no reason this manure can't be recycled into perfectly good feed... You should be really happy, 'cause this pretty much takes care of the 'organic waste' problem... Feed the animals shit and it gets rid of the waste at the same time" (258). What Gale describes has been linked to mad cow disease. Though cows are naturally herbivores, the industry turns them into cannibals by feeding them meal ground from beef and beef bones (Ratzan 1998; Rampton and Stauber 1997).

At the same time that systems of meat production became increasingly mechanized, systems of consumption were deliberately engineered so as to create new markets for the global food industry. In *My American Wife!* Beef-Ex sets out to familiarize Japanese housewives with meat and how to cook it, in a culture based historically on different food sources. Their strategy is "'to develop a powerful synergy between the commercials and the documentary vehicles, to stimulate consumer purchase motivation.' In other words, the commercials were to bleed into the documentaries and the documentaries were to function as commercials" (41). The novel thus connects meat production with global consumption, including advertising designed to create and shape the needs and desires of individual consumers in national and global markets. Whether that consumer is a housewife buying meat for her household table (in Japan or the United States), or a pregnant woman pressing her doctor to prescribe DES as a "magic" pill to prevent miscarriage, individual/group consumption and production are inextricably linked through cultural ideas and images, through the discourse of the wonders of technological progress and improvement.

DES and the Search for Environmental Justice

This reframing of DES and meat as problems of technological production and consumption updates Haraway's cyborg. She argues that "high tech culture's challenge of existing categories and dualisms" is liberatory because these dualisms are challenged and because "it is not clear who makes and who is made in the relation between human and machine" (1991, 177). But using DES as a case study in technological culture makes clear the maker *and* the made. The pharmaceutical and livestock industries made DES, used and promoted it widely, made sure that regulatory agencies were ineffective at protecting the public interest, and altered women's and livestocks' bodies in terribly troubling and culturally complex ways. To embrace the cyborg and the hybrid as emblematic cultural figures and to reject notions of bodily and environmental purity does not mean that we can't have a *politics* and *ethics* of technologically polluted bodies. The accountability here lies squarely with corporate polluters, weak regulatory agencies, and the consumers who depend on existing structures of production and consumption. Thus, I turn to the field of environmental justice to consider DES as a contemporary parable.

The connections between DES, *My Year of Meats*, and environmental justice can be seen in the topical links between the issues portrayed in the book and examples of current environmental-justice activism, as well as in the expansion of definitions of environmentalism to include race, class, gender, and injustice frames — cultural questions in which narratives and stories play a central part. One link is encountered in the struggles of Native women in the Arctic to organize against their exposure to persistent organic pollutants (POPs). In arguing that bodies are "first environments," Native activists have linked reproductive health, environmental health, and cultural survival (Silliman et al. 2006). POPs are extremely toxic, long-lasting, chlorinated, organic chemicals that can travel far from their emission sources (often thousands of miles) and that accumulate in animals, ecosystems, and people. In the 1980s, scientists began to find high levels of toxic chemicals (pesticides, insecticides, fungicides, industrial chemicals, and combusted wastes) far from their sources of production. POPs were discovered in Arctic indigenous populations in the 1980s, when a Nunavik midwife in the Canadian North offered to collect local breast-milk samples. These samples were supposed to be control samples from a "clean" environment in a study of toxic breast-milk contamination. Researchers were astounded to find that Arctic indigenous women had POP concentrations in their breast-milk five to ten times greater than in southern Quebec, and among the highest recorded in the world (Downie and Fenge 2003). Suspected health effects include higher rates of infectious diseases and immune dysfunction, negative effects on neurobehavioural development (e.g., slowed growth rates and impeded behavioural and intellectual functioning), and negative effects on newborn height.

The POP example raises questions of harms and injustices and rights and responsibilities, both for Arctic communities and in the global realm. Women in the North struggled to decide whether to feed their babies contaminated breast milk, and communities debated whether and how to hunt traditional foods. Native women, in other words, were cyborgs and hybrids, much like DES animal/female bodies and polluted babies. Arctic women suffered the worst health effects of production processes that their culture did not create or condone. Further research showed why and how these high rates of contamination were possible. First, POPs and organochlorines have unique properties. These contaminants are released (low estimates are 70 percent from the United States, 11 percent from Canada, and 5 percent from Mexico), then travel by atmospheric and ocean currents to the Arctic, where they enter the food chain. POPs are lipophilic ("fat loving"), which means they bioconcentrate in fatty tissues and biomagnify as they move up the food chain. Second, Arctic indigenous populations consume at the top of the food chain (especially marine mammals); thus, they absorb high levels of contaminants through their traditional diet, which includes the fat (blubber) of narwhal, walrus, and beluga whales.

Despite the unique technological and bodily problems that POPs present, there have been successful developments aimed at addressing this particular problem. In 1998 and 2001, two binding international treaties were signed: the 1998 Arhus Protocol and the Stockholm Convention. Both ban or limit the production, use, release of, and trade in especially toxic POPs; establish scientifically based criteria and specific procedures for establishing controls on additional POPs; and seek to prevent the development and commercial introduction of new POPs. These agreements would not have been negotiated without indigenous activism. Sheila Watt-Cloutier, the chair of the Inuit Circumpolar Conference, describes the indigenous negotiating position on POPs: "A poisoned Inuk child, a poisoned Arctic and a poisoned planet are all one in the same" (2003). Indigenous activists, especially midwives and female tribal leaders, privilege the environmentalist trope and discourse of interconnection and web of life, especially that of mother/child/culture/planet. They do so without sanctioning or celebrating the pollution that turns their bodies into paradigmatic sites of boundary breakdowns between cultures of global production and local consumption. Arctic activists are engaging in debates about the impact of technology on female/indigenous bodies and in the politics of what can and should be done in the face of the border crossings and cyborgian alterations that define our age.

Understanding DES through the framework of environmental justice allows us to understand and critique it as one of the two pathways that brought the environmental endocrine hypothesis to light (the other was the discovery that wildlife reproductive disorders are linked to chemical effluents and pesticides as well as to declines in sperm counts and quality; see Colborn, Dumanoski, and Peterson Myers 1996; Cadbury 1997). The environmental endocrine hypothesis is the emerging scientific consensus "that a diverse group of industrial and agricultural chemicals in contact with humans and wildlife have the capacity to mimic or obstruct hormone function ... by fooling it into accepting new instructions that distort the normal development of the organisms." Synthetic organic chemicals have been linked to two dozen human and animal disorders (Krimsky 2000). By understanding how technologies are intricately and inextricably linked with bodies and nature, we can understand how to criticize the problems that come with synthetic organic chemicals, rather than accepting their ubiquity as a hybrid, post-normal, and post-natural cyborgian state of the world.

Conclusion: Boundaries, Borders, and Blowback

As Aidan Davison (2001) suggests in *Technology and the Contested Meanings of Sustainability,* the contest over "the ideal of sustainability is at once a contest over the future of technology" (2001, 93). He describes human practice as the "drawing toward and into ourselves of worldly things: things living and nonliving, artefactual and ecological, human and non-human, earthly and heavenly" (ibid., 166). Technologically polluted bodies are a dark version of human and corporate practice, a fusion of things living (human bodies) and non-living (chemicals, plastics, pesticides), human and non-human, earthly and a synthetic hell created out of our culture's desire to engineer the natural through the technological. In telling truths that are fictions, in a narrative both documentary and fabulist, *My Year of Meats* and cyborgs, analyzed through a cultural framework, provocatively reframe DES as a story that continues to provide important insights into American values, corporations, and cultures. In one sense DES is "in the past," yet its negative effects on the health and well-being of second and third generations of those exposed to it remain relevant. DES is also a reminder of the ongoing dangers of chemical intervention for humans and animals, given that it was immediately replaced by other estrogenic growth-enhancing additives—and this in addition to widespread use of antibiotics (Dutton 1988).

Which returns us to the babies with whom we began. Clearly, the notion that babies are pure, natural, and outside culture is false. But perhaps the larger question is this: What kind of culture do we live in where babies are born with two hundred chemicals in their body, and where that level of pollution has become acceptable, even normalized? If pollution is ever-present, and if we are already hybridized and cyborgian as a result of our pollution exposure, what are the possibilities for cultural and environmental change? Knowledge, activism, and narrative make sense of technologically polluted bodies and perhaps can remediate these health and environmental problems. Narratives by environmentalists and writers such as Ruth Ozeki and Sandra Steingraber have focused on issues of toxicity, gender, and breast-milk contamination and are central to illuminating the stories of people—especially women—who have been intimately shaped by corporate pollution (Steingraber 2001). Thus, I want to end by reframing hybrids and boundary crossings through the environmentalist discourse of interconnection and the "web of life." In testimony about a DES daughter who died from cancer, her siblings recount that "one thing Betsy taught me is that the environment is a *complex web of life*; everything is *interwoven*... I think we need to use our best science with our best intuition to remember that everything is interwoven. If you make a disturbance, it can have consequences far beyond" (Susan Wood, talking about her sister Betsy, in Braun 2001, 60; italics added).

Beyond knowledge and activism, we need a cultural and political analysis and a vocabulary that makes sense of DES's roots and impacts. This analysis will reveal what happens when we look closely at the interconnections, interweavings, and webs: of race, gender, and nature as constructed categories; of the porous boundaries and borders between mother/child and human (woman) and non-human nature (animal); of production and consumption; of the environmental and the technological. Only through such intersectional and interdisciplinary analysis can we begin to consider where and how justice is at all possible in our complex age.

Works Cited

Adams, C. 2000. *The Sexual Politics of Meat: A Feminist–Vegetarian Critical Theory.* New York: Continuum.

Adamson, J., M.M. Evans, and R. Stein, eds. 2002. *The Environmental Justice Reader: Politics, Poetics, and Pedagogy.* Tucson: University of Arizona Press.

Apfel, R.J., and S. Fisher. 1984. *To Do No Harm: DES and the Dilemmas of Modern Medicine.* New Haven, CT: Yale University Press.

Braun, M. 2001. *DES Stories: Faces and Voices of People Exposed to Diethystilbestrol.* Rochester, NY: Visual Studies Workshop.

Buh, J. Blu. 2004. *The Fire Ant Wars.* Chicago: University of Chicago Press.

Cadbury, D. 1997. *The Feminization of Nature: Our Future at Risk.* London: Hamish Hamilton.

Carson, R. 1962. [2002.] *Silent Spring.* Boston: Houghton Mifflin.

Colborn, T., D. Dumanoski, and J. Peterson Myers. 1996. *Our Stolen Future: Are We Threatening Our Fertility, Intelligence, and Survival? A Scientific Detective Story.* New York: Dutton.

Cronon, W. 1996. "The Trouble with Wilderness: or, Getting Back to the Wrong Nature." In *Uncommon Ground: Rethinking the Human Place in Nature,* ed. W. Cronon. New York: Norton. 69–90.

Davidson, D. 2003. "Woe the Women: DES, Mothers, and Daughters." In *Gender, Identity, and Reproduction: Social Perspectives,* ed. S. Earle and G. Letherby. London: Palgrave Macmillan.

Davison, A. 2001. *Technology and the Contested Meanings of Sustainability.* Albany, NY: SUNY Press.

DES Action. http://www.desaction.org.

Downie, D.L., and T. Fenge, eds. 2003. *Northern Lights Against POPs: Combatting Toxic Threats in the Arctic.* Montreal and Kingston: McGill–Queen's University Press.

Dumit, J., with S. Sensiper. 1998. "Living with the 'Truths' of DES: Toward an Anthropology of Facts." In *Cyborg Babies from Techno-Sex to Techno-Tots,* ed. R. Davis-Floyd and J. Dumit. New York: Routledge. 212–39.

Dutton, D. 1988. *Worse Than the Disease: Pitfalls of Medical Progress.* Cambridge: Cambridge University Press.

Ellul, J. 1964. *The Technology Society.* New York: Vintage.

EWG (Environmental Working Group). 2005. "Body Burden: The Pollution in Newborns." http://archive.ewg.org/reports/bodyburden2/execsumm.php.

Fenichell, S., and L.S. Charfoos. 1981. *Daughters at Risk: A Personal DES History.* New York: Doubleday.

Fiddes, N. 1991. *Meat a Natural Symbol.* London: Routledge.

Haraway, D. 1991. *Simians, Cyborgs, and Women: The Reinvention of Nature.* New York: Routledge.

Holmes, H.B., ed. 1992. *Issues in Reproductive Technology: An Anthology.* New York: Garland.

Horowitz, R. 2006. *Putting Meat on the American Table: Taste, Technology, Transformation.* Baltimore, MD: Johns Hopkins University Press.

———. 2003. "Meatpacking." In *Gender and Technology: A Reader,* ed. Nina Lerman, Ruth Oldenziel, and Arwen Mohun. Baltimore: Johns Hopkins University Press. 267–94.

Human Rights Watch. 2005. "Blood, Sweat, and Fear: Workers' Rights in U.S. Meat and Poultry Plants." http://www.hrw.org/en/reports/2005/01/24/blood-sweat-and-fear.

Krimsky, S. 2000. *Hormonal Chaos: The Scientific and Social Origins of the Environmental Endocrine Hypothesis.* Baltimore, MD: Johns Hopkins University Press.

Lublin, N. 1998. *Pandora's Box: Feminism Confronts Reproductive Technology.* Lanham: Rowman and Littlefield.

Marcus, A.I. 1994. *Cancer from Beef: DES, Federal Food Regulation, and Consumer Confidence.* Baltimore, MD: Johns Hopkins University Press.

Merchant, C. 1999. "Reinventing Eden: Western Culture as a Recovery Narrative." In *Uncommon Ground: Rethinking the Human Place in Nature,* ed. W. Cronon. New York: Norton. 132–59.

Orenberg, C. 1981. *DES: The Complete Story.* New York: St. Martin's.

Ozeki, R. 1998. *My Year of Meats.* New York: Viking.

Rampton, S., and J. Stauber. 1997. *Mad Cow USA: Could the Nightmare Happen Here?* Monroe, ME: Common Courage.

Ratzan, S., ed. 1998. *Mad Cow Crisis: Health and the Public Good.* New York: NYU Press.

Raymond, J. 1994. *Women as Wombs: Reproductive Technologies and the Battle over Women's Freedom.* Melbourne: Spinifex.

Seaman, B. 2003. *The Greatest Experiment Ever Performed on Women: Exploding the Estrogen Myth.* New York: Hyperion.

Seaman, B., and G. Seaman. 1977. *Women and the Crisis in Sex Hormones.* New York: Bantam.

Serlin, D. 2004. *Replaceable You: Engineering the Body in Postwar America.* Chicago: University of Chicago Press.

Silliman, J., M. Gerber Fried, L. Ross, and E.R. Gutierrez, eds. *Undivided Rights: Women of Color Organize for Reproductive Justice.* Cambridge, MA: South End.

Spence, M.D. 1999. *Dispossessing the Wilderness: Indian Removal and the Making of the National Parks.* New York: Oxford University Press.

Stein, R., ed. 2004. *New Perspectives on Environmental Justice: Gender, Sexuality, and Activism.* New Brunswick, NJ: Rutgers University Press.

Steingraber, S. 2001. *Having Faith: An Ecologist's Journey to Motherhood.* Cambridge: Perseus.

Watt-Cloutier, S. 2003. "The Inuit Journey to a POPs-Free World." In *Northern Lights against POPs: Combatting Toxic Threats in the Arctic*, ed. D.L. Downie and T. Fenge. Montreal and Kingston: McGill–Queen's University Press. 256–72.

Chapter Six

Critical Mass: How Built Bodies Can Help Forge Environmental Futures
Fletcher Linder

This chapter explores how the built body can help forge environmental futures and is based on fifteen years of anthropological research on body-building in the United States. In detailing the material and subjective dimensions of bodybuilding, I reaffirm what Haraway (1991, 1995) identified as the cyborg ontological state of many late-modern bodies, and I illustrate how bodybuilding practice explicitly recognizes the metabolisms of late-modern human life, including how humans are entangled with Things. This recognition echoes a broader sensitivity to the technonatural realities of many late-modern human bodies and suggests that current imaginaries of the body transcend such oppositions as culture versus nature and human versus environment. I then suggest how academics invested in the creation of viable technonatural futures might step beyond critique to productively work within a culture that is both familiar with cyborgs and involved in body-based aesthetic craft. Possible strategies for cooperation include asking these questions: Toward what kinds of technonatural bodies might humans aspire? What metabolisms of production, consumption, and leisure are required to assist all humans in crafting such beings? How can these embedded human endeavours create just relations with Things? And how might intellectuals embody the kind of materially grounded engagements we hope to promote?

The Built Big Body as Cyborg Materiality

Bodybuilding is an aesthetic craft concerned with the development of the human physique. Media research suggests the number of American body-builders exceeds one million, only a small portion of whom ever compete in formal bodybuilding competitions (Linder 2007).

Bodybuilding began its modern existence in the swell of Greek Revivalism that spread across Europe, Great Britain, and the United States in the early nineteenth century. Much more than a simple assertion of ancient Greek aesthetic tastes, Greek Revivalism was in many ways a social movement that re-formed and reappropriated ancient Greek practices ranging from architecture to education. Roman and Hellenic statuary occupied a central place in Greek Revivalist movements, and the 1806 introduction of the Elgin Marbles in Britain created popular interest in "heroic" and muscular physiques. The famous marbles resonated with strongman shows which were popular during the early and mid-1800s, and provided a "high culture" referent to elevate the status of strongman entertainment (Chapman 1994).

By the late nineteenth century the physique dimension of Greek Revivalism had come into its own as a popular cultural concern. Newspaper coverage of Chicago's 1893 Columbian Exposition reported that fairgoers were fascinated with the muscular bodies on display. Physique and strength exhibitions by Eugene Sandow allegedly conspired with the fair's several dozen Hellenic statues to prompt many attending the fair to unfavourably compare their own physiques (ibid.).

Sandow, a Prussian-born Ziegfeld showman, and American-born Bernarr Macfadden were two of the most influential figures in this movement. Together they built on popular ancient Greek public imaginaries to develop what became known as the physical culture movement. This new social and aesthetic movement underlined Greek Revivalism's anti-modernist tendencies through the promotion of rigorous training designed to combat modern life's ill effects on body, mind, and spirit. Macfadden began publishing *Physical Culture* magazine in 1899; two years later he claimed more than 100,000 subscribers. He used the pages of *Physical Culture* to conduct what was perhaps the first modern physique contest for the "Best and Perfectly Developed Man and Woman." Contestants sent in photographs in the hope of being chosen for regional competitions and, ultimately, the 1903 finals in Madison Square Garden. Theodore Roosevelt, Jack London, Sir Author Conan Doyle, and Harvard professor

Dudley Sargent were among physical culture's more famous aficionados. By 1911 the physical culture movement had gained such high esteem that King George V appointed Sandow as Professor of Scientific and Physical Culture (ibid.).

During the early part of the twentieth century, bodybuilding in the United States lived in the shadows of Olympic weightlifting. Primarily under the direction of U.S. Olympic coach and American Athletic Union power broker Bob Hoffman, physique competitions were, at least until the 1940s, usually held late at night after weightlifting competitions had drawn to a close (Fair 1999).

In 1943, to draw more attention to physique competitions themselves, Montreal's Joe Weider, who had been publishing *Your Physique* magazine since 1940, formed the International Federation of Bodybuilders (IFBB) with his brother Ben (ibid.). This crucial shift has helped make body-building into a present-day international sport with virtually no ties to competitive weightlifting (Klein 1993; Linder 2007).

Bodybuilding is a craft primarily concerned with becoming *big*—a technical term used by bodybuilders to index bodies with high degrees of muscle development (called mass), muscle clarity and refinement (called definition), and an overall body shape that emphasizes shoulder width over waist size. Figure 6.1 illustrates how these components articulate in the qualitative notion of bigness.

Bodybuilders strive for bigness by following strict regimes of weight training, aerobics, diet, and nutritional supplementation—practices that

FIGURE 6.1
The Ideal Big Body

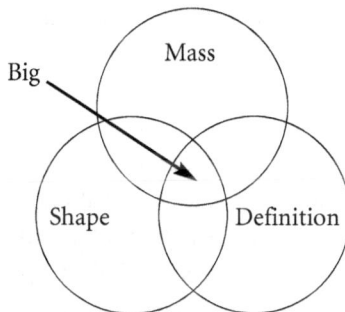

render the big body a cyborg. Here I use the term cyborg, following Haraway (1991), to refer to beings whose very life is made possible by complex interplays of biological and technological processes. Characteristics of cyborgs include an attraction to various social boundaries and their crossing (e.g., male versus female, human versus animal, organism versus machine), and a mythical power to reshape late modern futures beyond identity politics.

To understand the precise ways in which the cyborgian built body is infused with technology, it is instructive to contrast bodybuilding with its cousin, Olympic weightlifting. Olympic lifters use weights as objects in themselves, as goals to be reached by successfully lifting them. While numerous weight-training techniques are used in bodybuilding, the difference between bodybuilding weight training and Olympic lifting is that bodybuilders use weights as tools in a fundamentally aesthetic enterprise rather than as goals in themselves. This relationship is revealed in body-builder talk, as Lou, a former Mr. America, admonished me while I was training: "Before you deadlift, you have to decide what you're doing it for. Are you Olympic lifting, power lifting, or bodybuilding? If you're body-building you've got to cut the weight way down and use the lift to help you squeeze your traps, lats, and erectors [i.e., muscles in the back]. Use high rep[etition]s to work all the little muscles and squeeze blood into them. You want to make the lift as hard as you can and do it as many times as you can — 135 [pounds] should seem like 400 [pounds]."

This kind of labour, even for those accustomed to it, is extremely difficult. The difficulty of bodybuilding training lies not in lifting the weights but in the degree to which muscles, human intention, and weights interact to produce a good muscle "pump." This cooperative cyborgian endeavour of "getting pumped" refers to the increased blood present in the working muscles and in the human experience of muscle fullness that accompanies increased blood flow. A poor pump indicates that the human/technology cooperation has fallen short and that consequently the progressive task of bodybuilding itself has failed in its attempt at physical improvement.

Because pumps are difficult to achieve — especially so with novice bodybuilders — it is no surprise that capitalist enterprises have supplied legions of machines designed to help bodybuilders achieve physical transformation. This alliance between capital and physique development has been present since bodybuilding's Victorian-era emergence and remains central to the operation of all modern gyms (Green 1986; de la Peña 2003).

Weight-training regimens are too many to list, though most follow "split" routines that train only a portion of the body in any one workout. The type and number of exercises, as well as the number of exercise sets and repetitions used per set, also vary. Most bodybuilders use three to six exercises to train a large body part (e.g., the back), repeating the exercise movement between eight and thirty times (i.e., eight to thirty "reps") to complete one "set." Sets range between two and five per exercise, and total sets per major body part range between eight and twenty. Aerobic work is most often done with the aid of such gym equipment as stationary bikes, treadmills, and rowing machines. Rest days fall anywhere in the weekly schedule and are used to recuperate from weight training. These training regimes are technologies in themselves, in the sense that they are knowledges and strategies that, when well executed, build big bodies (Linder 2007).

Besides being users and products of iron and machine, bodybuilders use food, nutritional supplements, and ergogenic aids to sculpt their bodies. Bodybuilders shape their bodies not only by modulating the amount of calories they consume, but also by managing the intake of food's constituative elements — among these, proteins, carbohydrates, fats, vitamins, minerals, and water. In this sense, bodybuilders are masters of food discipline, often using reductionistic scientist language and reasoning to plan and rationalize consumption. Table 6.1 provides a partial list of common supplements, each of which is used to embody gym labour and other substantial investments of time, food, and money.

The big body comes into being through the modulation of nutritional throughputs, the management of human energy, and a consistent and mindful engagement with machinery. Big bodies require specific inspirations, aspirations, work, and material input. These combine to allow bodybuilders to surpass — indeed, to overcome — what they regard as a "natural body," a state they see in the bodies of non-bodybuilders, and the curb toward which they feel their own bodies slide. Bodybuilders dedicate years of practice to alter the physique's material existence, to manage the vicissitudes of a body that "is always one step ahead of you," as a veteran pro told me.

Perhaps ironically, the cyborg built body is subjectively lived as a product of technologies' work on an obdurate and inscrutable object, a material product tied to human activity but which is at some point impervious

TABLE 6.1
Supplements

Food-related

amino acids	boron	gama-oryzanol & ferulic acid
bovine colostrum	choline	canthaxanthin
carnitine	potassium	dessicated liver
chromium compounds	ginseng	glandular extracts
dibencozide	yohimbine	metabolic optimizer powders
medium chain triglycerides	lecithin	smilax compounds
multivitamins/minerals	niacin	protein powders
vanadyl sulfate	vitamin C	weight gain powders
vitamin E	creatine	weight loss powders

Steroids

Anadrol	Primobolan
Anavar	Sustanon
Deca-durabolin	Testosterone Cypionate
Dianabol	Testosterone Enanthate
Equipoise	Testosterone Propionate
Finaject	Testosterone Suspension
Methyltestosterone	Winstrol

Steroid esters

Acetate	Decanoate	Hexanoate	Phenylpropionate
Butyrate	Enanthate	Isocaproate	Propionate
Caproate	Formate	Nonanoate	Undecanoate
Cypionate	Heptanoate	Octanoate	Valerate

Other

Arimidex	Clomid	Human growth hormone
Clenbuterol	Cytadren	Ephedrine

to human intervention. The cyborg built body resists technohyphenation at the level of experience even as its material existence is inextricably tied to technology-infused practice. In this way the cyborg built body is profoundly technologically mediated while keeping one foot in what ethnoscientists might call "natural" ontological realms beyond technology's full reach. This complex cyborg subjectivity is further explained below.

Cyborg Subjectivity

When Haraway introduced cyborgs into postmodern theory in the early 1990s, there was little to suggest that the conditions producing cyborgs had penetrated subjective realms. In contrast, a cyborg ontological state extends into bodybuilding subjectivity and is quite self-consciously acknowledged by bodybuilders. For example, bodybuilders see themselves as crossing human/animal and human/machine boundaries. Alan Klein (1985, 1993) has noted these transgressions in the ways bodybuilders talk about themselves and their bodies. Arms become "pipes" and "guns," backs become coiled "anacondas," and hard training makes bodybuilders "machines" and "monsters." Bodybuilding training has women building "masculine" muscular bodies, and the physique contests have men participating in "beauty contests." Bodybuilders also clearly recognize how self, experience, and world intimately intertwine in their aesthetic pursuit. As Clarence, a top competitor, told me: "Discipline... that's what it's all about. If you don't have the discipline, you can't be a bodybuilder. Bodybuilders have to eat, drink, and sleep bodybuilding to be any good. Otherwise, we'd look and *be* just like everybody else." Clarence's physique, the materiality that defines his very being as a bodybuilder, depends on conscious and careful articulation with objects beyond himself. Here, he gives expression to bodybuilders' understanding of themselves as cyborgs — that is, as creatures owing their existence to their home environments located on the boundary between the culturally real places called *nature* and *technology*.

Bodybuilders, however, are not alone in their conscious cyborg subjectivity, as several late-modern contexts have produced similar understandings of human life. Biotechnology has played a major role in these transformations (Keller 2000; Martin 1994; Rabinow 1999) and has affected not only modern notions of reproduction (Clarke 1995; Morgan 2003; Rapp, Heath, and Taussig 2001; Rapp 1999) but also the horizons of health, and of life and death itself (Franklin and Lock 2003; Keller 1995; Lash 2001; Lock 2002a, 2000b, 2003).

This growth in technonatural understandings of human life signifies that cyborgism is no longer merely a descriptive tool used only in postmodern theory. At a strictly material level, cyborg ontological status grows out of a fundamental and inescapable human entanglement with things. Eating food, building dwellings, and even breathing are but a few such practices that link human survival with the material world beyond ourselves. Indeed, human life itself would be impossible without a constant entanglement with things.

What bodybuilders and other late-modern cyborgs do as a matter of course is explicitly point to this connected state. This explicit recognition of connectivity, however, should not be understood simply as being aware of "the true connected nature of things." This creation of cyborg subjectivity may also signal the logically inescapable consequence of what Bruno Latour (1993) has called the "purification processes" of modernity, processes of categorization that enable conceptual and material manipulation of the world. One unintended consequence of purification is the proliferation of hybrids. In simple terms, no human classification system works perfectly. So-called hybrids are but cases that do not fit neatly into existing cultural taxonomies, none of which can ever completely map the variation in the so-called natural world. The increasing cultural recognition of cyborg ontological forms, and the growing number of human subjectivities understood as cyborg, signal the crossing of such self-limiting categories as *human, nature, technology*, and *culture*.

Whatever its logical or historical necessity, cyborg discourse has entered popular culture in ways that render it enabling to human activity. By becoming cyborgs, bodybuilders develop big bodies. By joining forces with new reproductive technologies, would-be parents bolster their capacity to develop otherwise unlikely offspring. Even critics of cyborg life render the cyborg as enabling. Critics of reproductive technologies, for example, caution against the uncertainties inherent in "playing God" with human reproduction. Cyborgs are, in short, vehicles for human aspiration regardless of whether actions result in good or ill.

The cyborg's enabling power and ethical ambiguity are captured in bodybuilding supplement advertisements, such as the one shown in Figure 6.2.

In bodybuilding media, the cyborg is most often depicted as physically developed yet ethically uncultivated. As Figure 6.2's caption reads: "The future of prohormones is now. Use the best, be the best." While supplements can and do claim to help users become cyborg big, the cyborg's value as an agent of personal domination over others renders it without clear social value even as "MMSN team membership" apparently comes

FIGURE 6.2
*Flex Wheeler, Professional
Bodybuilder, Cyborg,
Technonatural Ally*

with supplement purchase. Interpersonally frightening yet potentially successful in competition, such images construct cyborgs as powerful agents of change and as frightful in their inability to positively direct that change. Progress and folly are equal possibilities for the cyborg, and it may be more than mere coincidence that the most famous cinematic cyborg is also a bodybuilder. Arnold Schwarzenegger, in his roles as the Terminator, embodies the powerful though ethically underdeveloped realities of cyborg living—technologically enabled, yet currently unable to focus that power toward a social good.

Joining Cyborgs, Dropping Critique
What I propose next is practical. I ask those involved in producing technonatural environmental discourses (e.g., those following Castree and Braun 1998, 2001; Castree 2006; White and Wilbert 2006) to enter into body-focused cyborg culture to help forge environmental futures. This section moves away from the practice of bodybuilding to provide intellectual and political justification for such an odd request, and in defending this tactic, I intentionally evoke religion. I do this neither to promote

religion nor merely to recall Haraway's (1991) use of blasphemy as the central metaphor motivating her introduction of the cyborg. I evoke the category of *religious experience* specifically to prompt readers to question how critique positions the critic in relation to unknowables — *religious experience* being one, and *futures of any sort* being another. Academic habits of critique have limited use in the pursuit of either, and the sooner academics develop alternative modes of engagement, the more capable they will be of shaping livable futures.

According to Bruno Latour, modern modes of intellectual critique were born of the Enlightenment. Their targets of inquiry are *matters of fact*, and they approach matters of fact by excavating the conditions within which facts are produced and gain legitimacy (Latour 2004, 231–36, 244).

Critiques based on matters of fact alone, however, appear to have rendered facts too unstable to support sound judgment and even cautious action. If facts are always "constructed," as is commonly and simplistically asserted, then facts can always be suspected of being mere products of ideological human work. Because of this suspicion, what historically has been a progressive strategy to expose ideological positions masquerading as scientific facts has now become an excuse for ideologues to disregard inconvenient facts, and for non-empirical skeptics to disregard facts per se.

Modern critics help destabilize facts by habitually using two broad and non-overlapping intellectual strategies, what Latour calls the "fact" and "fairy" positions (Latour 2004, 236–43). The fairy position permits the critic to explain away truth claims he holds suspect, while the fact position allows the critic to leave unscathed facts he holds dear.

These abstractions gain traction when we consider a specific case, one drawn from my home discipline of anthropology. In *Theories of Primitive Religion* (1982 [1965]), Evans-Pritchard noted how common it was for anthropologists to reproduce in studies of primitive religion their own relationship to religion. Even as theorists of primitive religion had religious backgrounds, the overwhelming majority constructed their models of religion at a time when they were either agnostic or atheistic (Douglas 1980, ix). Evans-Pritchard suspected this correlation explained the tendency he saw in all dominant anthropological theories of religion. Each theory — be it rooted in psychological, biological, or social modes of explanation — constructed primitive religion as an *illusion,* and "if primitive religion could be explained away as an intellectual aberration, as a mirage induced by emotional stress, or by its social function, it was implied that the [so-called] higher religions could be discredited and disposed of in the same way" (Evans-Pritchard 1982, 15).

This fairy position, this all-too-easy dismissal, clearly bothered Evans-Pritchard, perhaps in part because he had been since 1944 a practising Catholic (Douglas 1980, 43; Evans-Pritchard 1973, 37). Evans-Pritchard concludes his analysis of critique's negative influence on the study of religion by evoking qualitative dimensions of aesthetic experience, pointing to the life-denying dangers of fairy logic (1982, 121):

> As far as a study of religion as a factor in social life is concerned, it may make little difference whether the anthropologist is a theist or atheist, since in either case he can only take into account only what he can observe. But if [a theist or atheist] attempts to go further than this, each must pursue a different path. The non-believer seeks for some theory... which will explain the illusion; and... there is too much danger that the non-believer will talk of religion as a blind man might of colours, or one totally devoid of ear, of a beautiful musical composition.

To help avoid the debunking tendencies associated with modern critique, Latour (2004) suggests developing new critical tools, ones focused on *matters of concern*. A focus on matters of concern does not leave facts aside, but subsumes them in a *stubbornly realist attitude*, an attitude that takes *Things* as its objects of study (2004, 231). Things here stand in contrast to mere objects, as Things require the work of many participants to gather, assemble, and maintain their existence (ibid., 246). While a focus on Things is nothing new, especially in science studies; the novelty in what Latour proposes lies his insistence on a *productive* role for the critic.

Latour (2004, 247–48) looks to nuclear physics for a metaphor to tie together his designs for the new critic. A neutron entering a pile of "subcritical" atomic mass will simply fade away, while a neutron entering a "supercritical" pile will initiate a reaction releasing the pile's potential energy. The energy released by the neutron is determined by the mass and attitude of the medium into which it is inserted. The old critic, the one who simply negates, refutes, and takes away, lacks mass and the appropriate attitude to release stored energy. Latour's new critic is "supercritical" in that it has the characteristics to be *acted upon* and thus generate more energy than it takes in.

The following and final section provides an example of what it might mean to practise supercritical intellectual labour in the context of unknowables. It suggests ways to productively engage with the cyborg in the hope of forging viable environmental futures in technonatural times. The political component of this supercritical work is explicit in its awareness of popular sensibilities and is both committed to democratic deliberation and agnostic about the potential results of such deliberative processes.

Using Body-Conscious Metabolic Matters of Concern to Generate Environmental Futures

To summarize thus far, the historical emergence of the cyborg suggests a social recognition of the interconnected nature of life, the recognition that cyborg forms enable human agency, and a concomitant concern with the ethical outcomes of human agency enhanced by cyborg power.

This popular rendering of human life is reminiscent of various metabolic metaphors popular during the nineteenth century, many of which are evident in the broad range of intellectual work from Spencer to Marx (Castree and Braun 1998; Schmidt 1971; Swyngedouw 1999, 2006). As it was used in nineteenth-century European contexts, metabolism provided a body-centred metaphor within which intellectuals could describe and act in a world of human/non-human interaction and socioenvironmental co-evolution. For Marx, notions of metabolism helped him stress the dialectical relationship between human labour and all Things, including labour's role in shaping human social formations, ideologies, and subjectivities. Perhaps, given the proper materialist emphasis, contemporary academics can bring out the political potential of present-day cyborgs in similar fashion.

As Raymond Williams (1999) has noted, ways of understanding human existence in capitalist contexts are marked not by their tendency to be "too materialistic," as is often contended, but rather by their tendency to be not materialist enough, owing in large part to ideologies that downplay the constitutive power of material conditions. In order to combat this ideological tendency, I suggest activating the widespread critical mass of body-focused cyborgs to build environmental futures that *start from the individual and spread outward*, moving from material bodies to bodies politic. I suggest building on this historical moment in popular metabolic understandings, and thoroughly politicizing popular metabolic imaginings *by engaging with* and *adding to* rather than critiquing, denying, and otherwise taking away.

Below are three concrete steps to take toward forging environmental futures in an increasingly intimate partnership with cyborgs, heeding Haraway's (1992, 299) earlier message to academics: "Lives are built; so we had best become good craftspeople with the other worldly actants in the story."

Imagine and Observe

Imagine a social formation with fewer divisions between academic and popular aspirations. Observe popular culture focused on bodily pleasures and body aesthetics, and a proliferation of material designed to facilitate both. Picture popular cultural contexts already familiar with such notions as "technonatural" and more than willing to struggle with and debate the ethically ambiguous futures generated by cyborg life in technonatural times. Embryonic research. Artificial hearts. A hundred thousand more Dolly the sheeps, and this time for mutton. SUVs, hybrids, and flex-fuel automobiles move our bodies and race the road to attract consumer affection. Which commodity will win consumers' hearts and dollars in a time when the technonatural realities of gasoline-powered human mobility are as clear as ever? When international relations, corporate greed, and environmental policy seem to have a hard-wired connection to the price per gallon displayed on the gas pump?

Imagine metabolism as an organizing concept around which both popular and academic agents can, à la Laclau and Mouffe (1985), participate in democratic processes that enlist divergent social positions and passions. Grasp how "metabolism" places human bodies in an analogous relationship with the vital processes of all Things requiring throughputs to live, including urban landscapes — seemingly the most "unnatural" of spaces (Swyngedouw 1996, 2006) and home to half the world's population. Bring to mind a multitude of bodyscapes and social imaginaries that claim all animate beings consume to live. Envision how patterns of production and consumption lay the existential foundations for being. Picture humans caught up in working through the ethics of living lives consciously understood as inextricably bound with Things, both animate and not, and that human action and even modest forms of human labour help shape relationships between Things. Imagine social contexts wherein even the labour of the purchase can assume ethical weight. What is the production chain that put these grapes in my market? How might using such long-distance carriers as Working Assets connect my communication patterns to political movements? How will these kitchen-cleaning agents affect my septic tank and groundwater?

Observe the present in all its metabolic materiality and imagine entering the fray alongside bodybuilders, consciously connecting with and encouraging further metabolic imaginings, all in the service of quality development.

Connect and Encourage

Now go further. Do more than image, picture, perceive. Step out of the comfortable distance that critique provides. Or call on other modes of critique, ones that, like the artist/art critic (Williams 1983) or player/coach, works alongside the actor to improve the products of human labour. Engage in a more material fashion to connect with and encourage more politically developed metabolic cultural imaginings.

Connect with cyborgs and popular bodyscapes in the hope of coming to grips with what Stuart Hall (1992, 278) has called the "dirtiness" of theoretical work — the ways in which the semiotic game of intellectuals hits the ground of political practice. Though in many ways accurate, it may not be enough to argue, for example, that forces of capital are responsible for many of the environmental troubles taking us into the future. These positions often place environmental culpability solely at the feet of some abstraction, such as, "capitalism" or "the economy." A more likely agent of social change may involve the development of the more general Marxist sensibility toward the metabolisms of human life and the place of human labour in constructing those metabolisms.

This materialist ontological stance is already widespread enough to allow the political consequences to be drawn out in everyday life and popular imaginings. I have in mind promulgating in academic and popular settings a variety of body-based matters of concern, among them: Toward what kinds of technonatural bodies might humans aspire? What metabolisms of production, consumption, and leisure are required to assist all humans in crafting such beings? How can these embedded human endeavours create more just relations with Things?

These concerns keep individual lives focused on materiality and locate ethics in the outcomes of specific human actions. In working through these concerns, the heroes and villains will not likely take the form of capitalism, socialism, or even power relations, but will likely emerge from the realm of human activity itself, whether the action is concerned with imagining desirable future body forms or producing, consuming, and working toward these aesthetic goals.

Craft and Embody

Go further still. Craft, embody, and model. Live reflectively in the most intimate of environments, our own bodies. Environmental futures reach both outward to the world and inward to the self. Craft your own most intimate materiality.

Come full circle and look to built bodies as examples of how the metabolic understandings of human life play out in an actual human life. Try to occupy your body as self-consciously as bodybuilders do, with a thoroughly materialist understanding of our connection to Things — including people — around us. Recognize your role in shaping those connections. If we cannot live the metabolisms so prominent in our theoretical and political convictions, then why should we expect those around us to take our convictions seriously?

Whereas early forms of wage labour worked to separate the self from its activities, late capitalism has created technologies that intrude on and reconfigure the alienated and insular self. Indeed, the material conditions of late capitalism have created both the technonatural contexts and the body-focused cyborgs that live within them. Shouldn't we academics fully immerse ourselves in this world, if only to know what it feels like to live as much in our bodies as in our minds? Then, perhaps, body craft can assume an intellectually respectable reputation as a potent and metabolically oriented form of human activity. And perhaps only then can humans fully cultivate the aesthetic depth necessary to create broad-reaching techniques to care for, develop, and beautify our bodies and ourselves — those Things upon which our very lives depend.

Acknowledgments

Portions of this research have been supported by UNC Chapel Hill, James Madison University, and the Pugh Endowment. Much appreciation to Jennifer Coffman, Damian White, Chris Wilbert, Judith Farquhar, Christine Gearheart, and Frank Zane.

Works Cited

Castree, N. 2006. "A Congress of the World." *Science as Culture* 15, no. 2: 159–70.

Castree, N., and B. Braun. 1998. "The Construction of Nature and the Nature of Construction." In *Remaking Reality: Nature at the Millennium,* ed. B. Braun and N. Castree. New York: Routledge. 3–42.

———, eds. 2001. *Social Nature: Theory, Practice, and Politics.* Oxford: Blackwell.

Chapman, D. 1994. *Sandow the Magnificent: Eugene Sandow and the Beginnings of Bodybuilding.* Champaign: University of Illinois Press.

Clarke, A. 1995. "Modernity, Postmodernity, and Reproductive Processes, c. 1890–1990." In *The Cyborg Handbook,* ed. C. Gray, H. Figueroa-Sarriera, and S. Mentor. New York: Routledge.

de la Peña, C.T. 2003. *The Body Electric.* New York: New York University Press.

Douglas, M. 1980. *Edward Evans-Pritchard.* New York: Viking.

Evans-Pritchard, E.E. 1982 [1965]. *Theories in Primitive Religion.* Oxford: Oxford University Press.

———. 1973. "Fragment of an Autobiography." *New Blackfriars* 54: 35–37.

Fair, J. 1999. *Muscletown USA: Bob Hoffman and the Manly Culture of York Barbell.* State College: Penn State University Press.

Franklin, S., and M. Lock. 2003. *Remaking Life and Death: Toward an Anthropology of the Biosciences.* Santa Fe, NM: School of American Research.

Green, H. 1986. *Fit for America.* New York: Random House.

Hall, S. 1992. "Cultural Studies and Its Theoretical Legacies." In *Cultural Studies,* ed. L. Grossberg, C. Nelson, and P. Treichler. *Cultural Studies.* New York: Routledge. 277–94.

Haraway, D. 1995. "Cyborgs and Symbionts: Living Together in the New World Order." In *The Cyborg Handbook,* ed. C. Gray, H. Figueroa-Sarriera, and S. Mentor. New York: Routledge. xi–xx.

———. 1992. "The Promise of Monsters: A Regenerative Politics of Inappropriate/d Others." In *Cultural Studies,* ed. L. Grossberg, C. Nelson, and P. Treichler. New York: Routledge. 295–337.

———. 1991. *Simians, Cyborgs, and Women.* London: Free Association.

Keller, E.F. 2000. *The Century of the Gene.* Cambridge, MA: Harvard University Press.

———. 1995. *Refiguring Life: Metaphors of Twentieth-Century Biology.* New York: Columbia University Press.

Klein, A. 1993. *Little Big Men.* Albany, NY: SUNY Press.

———. 1985. "Muscle Manor: The Use of Sport Metaphor and History in Sport Sociology." *Journal of Sport and Social Issues* 9, no. 1: 68–75.

Laclau, E., and C. Mouffe. 1985. *Hegemony and Socialist Strategy.* London: Verso.

Lash, S. 2001. "Technological Forms of Life." *Theory, Culture, and Society* 18, no. 1: 105–20.

Latour, B. 2004. "Why Has Critique Run Out of Steam? From Matters of Fact to Matters of Concern." *Critical Inquiry* 30: 225–48.

———. 1993. *We Have Never Been Modern*. Trans. C. Porter. Cambridge, MA: Harvard University Press.

Linder, F. 2007. "Life as Art, and Seeing the Promise of Big Bodies." *American Ethnologist* 34, no. 3: 451–72.

Lock, M. 2003. "On Making Up the Good-as-Dead in a Utilitarian World." In *Remaking Life and Death: Toward an Anthropology of the Biosciences,* ed. S. Franklin and M. Lock. Santa Fe, NM: School of American Research. 165–92.

———. 2002a. "Utopias of Health, Eugenics, and Germline Engineering." In *New Horizons in Medical Anthropology,* ed. M. Nichter and M. Lock. London: Routledge. 239–66.

———. 2002b. *Twice Dead: Organ Transplants and the Reinvention of Death.* Berkeley: University of California Press.

Martin, E. 1994. *Flexible Bodies: Tracking Immunity in American Culture from the Days of Polio to the Age of AIDS.* Boston, MA: Beacon.

Morgan, L. 2003. "Embryo Tales." In *Remaking Life and Death: Toward an Anthropology of the Biosciences,* ed. S. Franklin and M. Lock. Santa Fe, NM: School of American Research. 261–91.

Rabinow, P. 1999. *French DNA: Trouble in Purgatory.* Princeton, NJ: Princeton University Press.

Rapp, R. 1999. *Testing Women, Testing the Fetus: The Social Impact of Amniocentesis.* New York: Routledge.

Rapp, R., D. Heath, and K.S. Taussig. 2001. "Genealogical Dis-ease: Where Hereditary Abnormality, Biomedical Explanation, and Family Responsibility Meet." In *Relative Matters: Reconfiguring Kinship Studies,* ed. S. Franklin and S. MacKinnon. Durham, NC: Duke University Press. 384–409.

Schmidt, A. 1971. *The Concept of Nature in Marx.* London: New Left.

Swyngedouw, E. 2006. "Circulations and Metabolisms: (Hybrid) Natures and (Cyborg) Cities." *Science as Culture* 15, no. 2: 105–21.

———. 1999. "Modernity and Hybridity: Nature, Regeneracionismo, and the Production of the Spanish Waterscape, 1890–1930." *Annals of the Association of American Geographers* 89, no. 3: 443–65.

———. 1996. "The City as a Hybrid: On Nature, Society, and Cyborg Urbanization." *Capitalism, Nature, Socialism* 7, no. 2: 65–80.

White, D., and C. Wilbert. 2006. "Introduction: Technonatural Time-Spaces." *Science as Culture* 15, no. 2: 95–104.

Chapter Seven

Living Between Nature and Technology: The Suburban Constitution of Environmentalism in Australia
Aidan Davison

A central theme of the present book is that Western environmentalists have found it increasingly difficult to orient political practice and scholarly critique in the face of technonatural complexity. The ways that environmentalists represent nature as an autonomous realm beyond culture are well documented (e.g., Braun and Castree 1998; Cronon 1996; Latour 2004). Less understood are the ways they similarly represent technology as unfolding according to its own inherent designs. Reflecting the influence of technological determinism in Western societies, environmentalists are apt to portray technology as not only out of human control, but as in control of humanity.

Whereas questions of nature have united Western environmentalists, questions of technology have often polarized them (Davison 2001). Crudely put, the first wave of postwar environmentalism that took shape in the 1960s displays a technophobic or neo-Luddite sensibility. Associated with countercultural movements, neo-Luddite environmentalism emphasizes ecological limits to social development, sees nature pitted against artifice, and advocates smallness and simplicity as technological virtues. Though prominent still in wilderness-based movements (e.g., McKibben 2003; Sessions 2006), the influence of this sensibility has ebbed since the 1980s as a new wave of environmentalism has gathered strength. Displaying a technophilic or Promethean sensibility, this second wave of postwar environmentalism is associated with the professionalizing of environmental knowledge and has strong links to modern institutions. Formed around the language of sustainable development, this Promethean sensibility draws momentum from a perceived convergence of capitalist and environmentalist objectives (e.g., Hawken et al. 1999). Taking technology to be nature's salvation, Promethean environmentalism has as its main pursuit eco-efficiency, a concept minted by the new breed of corporate environmentalist. Indeed, some Promethean environmentalists regard nature itself as technology (Davison 2008a).

With the entry of environmental issues into mainstream professions, institutions, and political debate over the past twenty-five years, environmental movements and environmentalist identities have diversified and fragmented. Environmentalism is increasingly difficult to define, and environmentalists have been caught up in internecine conflicts, especially in the field of environmental philosophy (Hay 2002). The clash of neo-Luddite and Promethean sensibilities has been a common feature of environmentalism since the 1980s, though it is more usually presented as a conflict between radical and reformist philosophies or between fundamentalist and realist politics. Less obvious has been the overlap between these two sensibilities in the form of a shared debt to dualistic accounts of nature and deterministic accounts of technology. This shared debt has played an important role in the creation of the informal networks and implicit alliances by which environmentalism has increasingly progressed as the variety of environmental movements and environmentalist identities has grown (Doyle 2000; Milton 2002). These important if weak and shifting alliances are to be found not just among different movements, but in the lives of environmentalists as they shuttle between different environmental discourses and modes of activism. One of the more obvious examples of this multiplicity is the way Western environmentalists rely on scientific knowledge to establish that humanity is a part of nature, while relying on sensual experience to establish that nature is a sacred Other (Chaloupka 2008; Milton 2002).

Accepting that the contemporary reality of environmentalism is far less neat than the metaphor of waves allows, there are recent signs that Promethean environmentalism may be losing its momentum and shape. From debate among environmentalists about the "death" of environmentalism (Shellenberger and Nordhaus 2005; Schlosberg and Bomberg 2008) to analysis of internal contradictions in environmental movements (Chaloupka 2008; Latour 2004) to calls for a post-sustainability environmental paradigm (Blühdorn and Welsh 2007), there is growing recognition by environmentalists that some past concepts and political strategies are serving them poorly. In this context, the present book draws attention to how ideas of nature and technology have failed environmental movements. Discussion of technonatures belongs to an emerging wave of environmental politics that is gaining momentum from scholarly projects as different as post-humanist political ecology and non-equilibrium ecology, as well as from emerging social movements such as the environmental-justice movement. This is not to suggest that a technonatural sensibility is new to environmentalism. In the 1960s and 1970s the political theorist Murray Bookchin warned environmental movements against any deterministic embrace or rejection of technology (Davison 2001). At the same time, ecologists such as C.S. Holling were debunking narratives about the non-human balance and purity of nature, laying the groundwork for contemporary analysis of socio-ecological systems (Folke 2006). It *is* to suggest, however, that a technonatural sensibility is beginning to renegotiate the terms of nature and technology in ways that more faithfully represent the ambivalence of contemporary materiality, thereby opening up less self-defeating forms of environmentalist engagement with modernity.

The rise of a technonatural sensibility within environmentalism is nowhere more evident than in the interest taken in urban nature over the past decade. In keeping with the way this sensibility resists the division of academic labour between society and nature, interest in urban nature is evident in social studies (Desfor and Keil 2004; Heynen, Kaika, and Swygedouw 2006) as much as in ecology (McKinney 2008; Miller and Hobbs 2002). This interest stands against a history of neglect founded on the assumption that urban space and natural space are mutually exclusive (Williams 1973). Scientists in search of nature have headed out of the city, while urbanists have taken the city to be an entirely human artifact. Fields central to environmentalist scholarship, such as those of environmental history (Isenberg 2006) and environmental ethics (Light 2001), have until recently largely disregarded the environments in which the majority of Western environmentalists live.

In this chapter I argue that a capacity to make sense of technonatural complexity is vital to the development of an effective and self-aware environmentalist agenda for Australian cities. Though 85 percent of Australians live in and around six suburban conurbations, the idea of suburban nature continues to have the status of an oxymoron in environmentalist discourses in which the terms "wilderness" and "nature" are virtually synonymous. In the first section I present the modern suburb as an important site where ideas of nature and technology have been negotiated, before sketching out Australia's suburban history and introducing the anti-suburban lament of environmentalists. In the second section I draw on interviews with environmentalists living in the cities of Perth and Hobart to explore the relationship between ideas of nature and suburban experience. In the final section I discuss prospects for a technonatural politics of suburban life in Australia.

The Modern Suburb in Australia

In addressing the modern suburb as technonature, I challenge instrumentalist and determinist accounts of technology as well as dualist concepts of nature. The two-way theory and practice has been hidden in many histories of modernity by the assumption that practices result from the application of ideas that are themselves the product of pure, disembodied reason (Taylor 1989). Such accounts of modernity reduce the materiality of technology and nature to abstractions. This emptying of material experience produces incomplete accounts of how modern worlds are made, reproduced, and transformed, and how they are enmeshed in nature (Latour 1993). Reduction of technology to transparent tools has the paradoxical effect of encouraging determinism by hiding the ways technology is constituted by and is constitutive of political processes. Equally, reduction of nature to raw material for technology has the paradoxical effect of encouraging romanticized representations of sacred nature that mask their political origins. In what follows, then, I understand the suburb as a complex of practices integral to and active within the modern history of ideas of technology and nature. After sketching out the British origins of the modern suburb, I focus on Australia's suburban history.

The modern suburb emerged alongside Enlightenment reforms in eighteenth-century Britain. Beginning as the unplanned experiments of an emerging middle class of merchants and industrialists, the suburb offered sanctuary betwixt and between city and country. Sustained by urban economies and by rural imaginaries, suburbs were a product of and a reaction against the advance of scientific reason and technological mastery. Of particular importance was the way they melded industrial capital, Romantic aesthetics, sanitarian science, and Protestant privatism

(Davidoff and Hall 1987; Fishman 1987; Williams 1973). For example, suburbs combined Protestant and Romantic narratives about the spiritual innocence of women, children, and rural life, and about urban evil, with the opportunities and dangers that characterized the industrial city. As Fishman (1987, 71) explains:

> in suburbia the conquering bourgeoisie had chosen to re-create an invented version of the "feudal, patriarchal, idyllic" village environment it was destroying ... At the same time that bourgeois economic initiatives were swelling the metropolis and undermining the traditional balance between man and nature, this class was creating a private retreat that expressed tradition, domesticity, and union with nature.

The juxtaposition of the industrial city and picturesque suburb gave spatial form to the modern metaphysics that located nature, women, and the body in a dimension of reality separate from and subordinate to culture, men, and the mind. However, the suburb was as much a rhythm of movement as it was a spatial division. The boundary between industrial order and suburban idyll was selectively permeable. It enabled the male pioneers of progress to journey back and forth between the industrial city, in which moral progress was found in material achievement, and the chaste suburban home, in which the spirit was renewed in God's garden by the virtues of family life.

In sheltering tradition, sentiment, religion, and nature from the glare of reason, the early modern suburb was not a deviation from the project of technological progress. It was, rather, a means by which apparently untenable contradictions in this project were resolved in everyday experience. While the beneficiaries of instrumental reason may have paid for "their power with alienation from that over which they exercise their power" (Horkheimer and Adorno 1972, 9), the suburb offers an example of a practical strategy by which experience of union with an eloquent world was maintained. The bourgeois suburb gave expression to an aesthetic and spiritual reaction against the dehumanizing and denaturing character of modern technology, while containing and deactivating such critique in a private realm at the margins of society. While later suburban experiments, such as the early-twentieth-century garden-city experiments associated with Ebenezer Howard, represented more substantive forms of political critique, suburbs have continued to be an important mechanism in the progress of modernity by providing carefully circumscribed refuge from it.

The role of suburbs in venting disquiet about the modern project was especially important in the settlement of the New Worlds of Australasia and North America, lands that had been stripped of tradition and in which Enlightenment reforms thus encountered less resistance than in Britain (Gascoigne 2002). Indeed, by the time it gained independence from Britain in 1901, little more than a century after settlement, Australia was the world's most suburbanized nation (Davison 1995). Australia's first suburbs did not exist between city and country; rather, they belonged to a radically novel form of city, one that was suburban before it was urban. By the end of Australia's first suburban boom, that of the 1880s, Melbourne stretched over no less than 164,000 acres, at a density of three people per acre, while Sydney was home to four people per acre over 96,000 acres (Weber 1969, 139).

In Australia as in Britain, suburban retreat was linked to the advance of reason. Anti-urban, Romanticist distrust of the satanic mills of modernity was influential in Australia, as was a yearning for the innocence of Eden and Evangelical emphasis on the sanctity of private domestic life (Davison 1995). Settlers were faced with a wild, unfamiliar nature apparently free of culture. At the same time, the emerging modern order was thoroughly and rapidly altering the cultural experience of settlers. In practice, this modern order was often as unfamiliar as the wilderness over which it claimed dominion (Gascoigne 2002), and many sought suburban refuge between the worlds of nature and technology.

Unlike Britain's early suburbs, Australia's suburban ideal was not primarily allied to the middle class but belonged to a suburban peasantry. Disquiet about modernity was played out in Australia in pursuit of an essentially pragmatic vision of Eden. As Hogan (2003, 69) notes: "Australia's is a strictly transposed, retrospective and novel reaction to European and North American industrial city experience. Of far greater importance to the shaping of suburban consciousness in Australian cities is the experience of domestic natural ecosystems and the 'socialization' of these ecosystems through gardening, garden economies, stakeholder democracy, back-shed poiesis [production], network utilities, leisure pursuits, and so on."

Many of Australia's working-class suburban pioneers were born into rural life in Britain before suffering in urban slums and then escaping (or being shackled) to the "farthest suburb" of Britain's global empire (Davison 1995, 52). These origins go part way to explaining how the suburban idyll was inflected in Australia by quasi-rural encounters with nature as a domain of productive labour. Just as important, however, was the economic structure of Australia's late-nineteenth-century mercantile cities. These economies were characterized by cheaper land and higher wages than in late-nineteenth-century Britain or North America, leading to high levels of home ownership, but also by low levels of pubic revenue, which explained the woeful state of public infrastructure if not its absence altogether (Mullins 1981). By 1890, for example, when more than half of Australia's housing stock was owner-occupied (Butlin 1964), Melbourne, a city of almost half a million people, had no sewerage system. The last of the city's pan toilets (from which nightsoil was removed in buckets) was decommissioned only a century later. In these conditions thrived what Mullins (1981, 69) has called a widespread condition of "forced self-sufficiency." Suburban homes became vital sites of food, clothing, furniture and housing production, water collection, and waste disposal. As is often the case with the lives of the working class, documentation of the private labour of Australia's suburban peasantry is scarce. Consider, however, the finding of the 1881 census that 40 percent of households in one Melbourne suburb kept pigs, sheep, or cows; consider as well that, as late as the 1941 census, two-thirds of Melbourne's suburban households were producing at least part of their own food (Gaynor 2006, 19, 102).

The private autonomy of Australia's suburban peasantry weakened through the first half of the twentieth century alongside belated industrialization and improved infrastructure. With the second suburban boom, after the Second World War, the transformation of suburbs from spaces of production into spaces of consumption was accelerated by the formation of a web of mutually reinforcing relationships between industrial production and suburban consumption. This suburban-industrial complex combined cheap oil, land speculation, housing construction and finance, and systems of mass production to drive economic growth. In these conditions the majority of Australia's postwar population, swollen by a local baby boom and immigration from Europe, was to be found in suburbs. Bearing testament to the success of Australia's ideal of a householder democracy, 71 percent of the nation's housing stock in 1966 was owned or being bought by its occupants (Troy 2000, 719).

Though postwar suburbs were an overt driver of economic and technological modernization, they continued to be spaces in which nostalgia for nature and disquiet about modernity were embodied. As the practices of the suburban peasantry were dismantled, suburbs became increasingly significant sites of sentimental encounter with innocent nature—in the form, for instance, of non-utilitarian gardening and pet keeping and outdoor recreation in suburban parks and beaches. As the 1967 *Reader's Digest Complete Book of the Garden* explains, the goal of suburban gardeners was now to work "together with nature to develop a landscape of unaffected charm, one that epitomizes natural beauties" (cited in Timms 2006, 66). Backyard chicken runs, incinerators, and "dunnies" (privies) gave way to the demands of entertainment and enchantment. Much of the infrastructure now vital to the life of Australian suburbs was buried beneath swathes of lawn, rendering the technological production of the city largely invisible and apparently autonomous.

Given their mass congregation in suburbs during the first two-thirds of the twentieth century, it might be expected that suburbs figured prominently in the stories Australians told about themselves. However, even though they represented the single most monumental technological endeavour of the previous century, suburbs were almost invisible in representations of national identity and in discourse of technological progress. To the extent that suburbs were visible in Australian culture, it was in the form of intellectual and artistic scorn for suburban "half-worlds" (Boyd 1952, 6), a tradition with its origins in the late nineteenth century (Gilbert 1988). The legend that located Australian identity "out back," toward the continent's red centre, in remote sunburnt landscapes sparsely populated by self-sufficient bush pioneers, continued to exert powerful influence, though it had always accounted for the experience of only a minority of settlers (Devlin-Glass 1994). Living out this legend in coastal cities, generations of Australians retreated to the limits of the city so as to gaze out upon nature. Such quintessentially suburban retreat pushed these limits ever farther out, creating vast cities. Again and again, suburban pioneers found themselves all too quickly trapped behind expanding brick frontiers, swallowed whole by what has come to be known as suburban sprawl. A frequently self-defeating compromise between the allure of nature and the necessity of the city, suburban preoccupations seemed transparent, lacking any real substance in their own right.

As the postwar suburban boom slowed in the late 1960s, the neo-

Luddite wave of postwar environmentalism took shape, gaining momentum from the paradoxical power of anti-suburbanism in a suburban nation. In championing and fighting for the virtues of wilderness, the new defenders of nature turned their backs on the suburbs where most of them had grown up, and in which many continued to live (Hutton and Connors 1999; Pakulski and Tranter 2004). The experience of having childhood haunts at the suburban fringes destroyed by the next wave of bulldozers was commonplace, endowing some with lifelong antipathy for anything suburban (Lines 2006). This experience perhaps helps explain why, in environmentalist discourse, suburbs appear to grow as if by themselves, with the result that Australia's suburban condition is regarded as a form of mass social entrapment rather than a deliberate social project.

While wilderness-based environmentalism remains strong today in Australia, a second, Promethean wave of environmental discourse has been on the rise since the late 1980s. Focused around the concept of sustainable development, this second wave has seen environmental concerns routinized (Pakulski and Tranter 2004). Yet as they have been made routine, these concerns have been redefined in ways that have unsettled and marginalized the neo-Luddite sensibilities of earlier forms of environmentalism, leading to a decline in their social influence (Davison 2008b). On the one hand, "the environment" has been translated into political portfolios, laws, bureaucracies, curricula, professions, and markets; the idea of sustainability has become a catch-all; more than half of all Australians admit to being "a bit of a 'Greenie' at heart" (Morgan-Poll 2000); and practices such as recycling and water conservation have become commonsensical. On the other hand, the proportion of the population concerned about environmental problems declined steadily between 1992 and 2004, from 75 to 57 percent, and from 80 to 49 percent in the 18-to-24 age group (ABS 2004), with the proportion belonging to environmental groups moving between 2.5 and 5.5 percent (Pakulski and Tranter 2004, 228). By 1997 the politically conservative environment minister, Senator Robert Hill, could with reason assert that "the whole environment debate has changed...Everyone now is an environmentalist" (cited in Hutton and Connors 1999, 264).

A major achievement of Promethean environmentalism has been to raise the policy profile of questions of urban sustainability. Just as dismissive of Australia's suburban sprawl as neo-Luddite environmentalism, Promethean environmentalism advocates not retreat from the city to nature, but efforts to make the city more urban. The remedy to suburban alienation, in this discourse, is to improve the eco-efficiency of urban machinery through market-led urban consolidation. In keeping with the sustainable-development goal of simultaneously achieving economic, environmental, and social outcomes, this remedy promises reduced environmental impacts through reduced resource use and waste production, market efficiencies through reduced state involvement in urban development, and improved social cohesion through increases in social capital (Newman and Kenworthy 1999). This promise has proven attractive; several of Australia's major cities, including Melbourne and Sydney, have recently adopted metropolitan plans embodying this vision of sustainable development, and the federal government instituted a parliamentary inquiry into urban sustainability in 2003. Revealing how tightly Australian urban-sustainability discourse is allied to Promethean hopes for technological salvation, none of the thirty-two recommendations of the report of this inquiry draw from the concepts of nature and ecology (Parliament of Australia 2005). These recommendations are entirely populated by humans and their tools, with non-humans of any sort failing to rate a mention. Yet despite the attractiveness of the Promethean promise of sustainable development, there is little evidence that the market-led urban consolidation that has already taken place in Australian cities over the past twenty years has significantly improved their sustainability (Davison 2006; Gleeson 2006).

As suggested at the beginning of this chapter, neo-Luddite and Promethean sensibilities overlap in complex ways in contemporary environmentalism, despite their apparent polarity. In Australia, neo-Luddite desires for wilderness are often allied with Promethean plans to re-create cities as efficient machines, through the ability of environmentalist discourse to switch back and forth from matters of nature to matters of technology. The half-worlds in which so many Australians live have confounded this discourse. Until recently, suburbs elicited from environmentalism neither the inspiration of nature nor the promise of technology, with the result that the question of suburban sustainability has rarely been asked. Given that many Australian environmentalists live in suburbs, it is equally remarkable that the question of how environmentalists make sense of their own suburban lives has attracted little attention. It is to this question that I now turn.

Bringing Environmentalism Home

In what follows I draw from thirty semi-structured hour-long interviews with active members of the Wilderness Society and of local branches of the national Green Party living in Hobart (pop. 220,000) and Perth (pop. 1,500,000), the capital cities of Tasmania and Western Australia, respectively. The interviews were conducted in 2004 as part of a larger study detailed elsewhere (Davison 2008b). I treat the sample as a heterogeneous whole that combines a wilderness-based activist organization and an environmental organization with several state and federal political representatives whose platform stretches from issues of wilderness preservation to those of social justice. A significant proportion of the participants are members of both organizations. My aim is to offer a brief yet suggestive inquiry into the relationship between suburban experience and environmentalist representations of nature. In keeping with the demographic profiles of Australian environmental movements (Pakulski and Tranter 2004), the sample strongly represented tertiary qualifications (90 percent), Anglo-Australian heritage (83 percent), women (66 percent), and the postwar generation (52 percent). Pseudonyms are used.

The interviews encompassed considerable diversity, with participants contextualizing and qualifying their comments about nature. Only one participant in five agreed with the proposition that "city life is less environmentally sustainable than rural life." Indeed, the finding that only one in six disagreed with the proposition that "high-density housing is more ecologically sustainable than low-density housing" suggests that the connection drawn in sustainable-development discourse between compact cities and sustainability has become widely accepted by environmentalists. Consistent with this, it emerged that most participants considered suburban development to be the principal threat to the sustainability of Australian cities. Janina (Perth), for example, contended that "within our sprawling, sprawling cities we are just destroying everything in our path," while Sandra, from the more "boutique" capital city of Hobart, reflected that "maybe that's why I came to Tasmania — you don't get that urban sprawl, but when you go to the mainland and you see the spread, that's horrible." Similarly, and despite that Hobart has a lower population density than that of Australia's largest cities, another of the Hobart participants, Mike, contended that "if we are going to come to terms with our impact on the Australian environment, somehow we have got to stop the massive sprawl of those five cities" (i.e., the largest "mainland" capital cities). This tendency to direct criticisms at distant cities was common.

Environmentalist disaffection with suburbs rehearsed older anti-suburban themes about social sterility. So, for example, when asked whether she considered herself to be suburban, Mandi (Hobart) replied: "Oh God, no!... Well, I lived in the suburbs for about twelve months, I lived in Doncaster [Melbourne], and I found it really oppressive, the time that I spent there, not because of the people, but because of that sort of Saturday morning 'we get up and we wash the car and mow the lawns'... There seems to be a lack of expression in the suburbs as compared from either really inner-city or really out of the city."

The theme of lawns was common. Participants linked the social with the ecological shortcomings of suburbs. Reflecting on her alienation from neighbours in a northern suburb of Perth, one recent immigrant from Europe, Nadia, observed: "You can just see it, the sort of stuff they buy for their kids. It's very, very materialistic... and spring lawn everywhere."

Besides signifying materialism, lawns were for many an archetypical embodiment of "unnatural nature." Discussions about lawns carried with them stories about what it means to belong in distinctively Australian nature. Kylie (Perth) reflected: "I think I retain enough of my Anglo background to probably deeply connect with rolling green, but... I've kind of beaten that out of myself because I know of the environmental impact of maintaining lawns." Here Kylie was making explicit a tension between unreflective environmental experience and scientific knowledge—a tension implicit in many interviews. Unlike most participants, Kylie was aware of irony in her anti-suburbanism: "When I was younger I was incredibly dismissive of suburbia and would have done anything to avoid living in such a soulless... sort of place. [*Interviewer: And now?*] P'raps it is an age thing, because I probably still do have that feeling about living in a brand new subdivision, y'know, out in the fringes of Perth. Maybe I've transferred my feelings about the death of suburbia or something to those fringes. And because the suburb where I live is probably more that middle-ring... and its 1950s, 1960s, maybe it just sort of feels just slightly more embedded again."

As many as two-thirds of the interviewees lived, as did Kylie, in a detached house with a private garden in a middle-ring, middle-class suburb on a block ranging from one-eighth to one-third of an acre. Only three participants lived in an apartment, with one of these subsequently moving into a detached house and another aspiring to own a house suitable for keeping a dog. Consistent with the high proportion of tertiary-educated professionals, more than two-thirds of the sample owned or were purchasing a home. Yet despite their suburban location and commitment to private home ownership, there was a noticeable lack of positive sentiment about suburban environments and suburban aspirations. When I asked in what kind of environment they would most like to live, only 15 percent nominated a suburban environment, with 33, 22, and 19 percent electing rural/remote, peri-urban, and inner-urban environments, respectively.

It might be expected, then, that many participants would be unhappy with their home environment. Yet I found that, regardless of the apparent similarity of their own everyday life with much of what they criticized, few participants identified themselves with generalizations and stereotypes about sprawl and suburbanites. When speaking in specific terms about their homes and neighbourhoods, rather than in terms of general categories, many expressed strong attachment to their everyday environment. In particular, those living in Hobart—a small city squeezed long and thin between a mountain and an estuary—almost unanimously conveyed the sense that their home environment provided enriching encounters with nature and welcome access to the city. As Paul (Hobart) put it, they were living in "the best of both worlds." Similarly, Claire (Hobart) considered herself "really fortunate to have a good view, so you've got that sense of space and feeling a part of the city as well."

While several participants remarked disapprovingly on the growing size of suburban dwellings, desire for an everyday sense of space was strong for many. When narrated in material terms, this sense of space translated into intimate encounters with more-than-human presence and to freedom from unwanted social presence in the city. Asked about the views provided by her mountainside suburban home, Helen (Hobart) exclaimed: "Oh, I love it! Yes… when I look out my window of a morning from my bed, I watch the sun. I see the sun come up. And the sense of space, looking out—that sense of space is fantastic." Recalling a golf course he frequented as a child in postwar suburban Melbourne, Greg (Hobart) was equally emphatic: "Oh, the space. There was just so much space. And there were trees, islands of trees everywhere, and there were lots of areas of bush, particularly one on the edge of the golf links which was totally untouched, it was just bush."

For Henry (Perth), as for several others, the experience of a sense of space met a need for "independence and seclusion and space from other people... I'm always doing gardening here, and it's sort of like the space around me, y'know? I like the natural environment, the no pollution part of it." Hendrika contrasted her experience of Australia's low-density cities with her childhood in urban Europe: "Man has ruined it [Europe]. There are too many people. It drives you mad... The people have ruined it for themselves, and that's why there is such an appreciation of places like Tasmania and Australia in general by Europeans, because [of] the space, the influence of the space that does you good."

Discussion of the city's soundscapes was common. Participants elaborated on the link between an everyday sense of space in nature and freedom from social intrusion. Barbara (Hobart) liked "space to think without noise, and I like to think that I can have some unobserved space in my residence." Simon's decision to buy a detached house with a good-sized garden near suburban bushland in Perth was influenced by the fact that he "can't *stand* noise, particularly other people's noise... well, *only* other people's noise! I can understand noises that don't emanate from people."

When the participants talked about their home and neighbourhood, terms such as *space* and *peace* stood in place for terms such as *nature* and *ecology*—ideas most commonly presented as mutually exclusive with the idea of the city. Interviewees made no use of the concept of urban nature, and most were convinced that real (i.e., pure) nature existed only beyond the city's reach. Nonetheless, there seemed broad agreement that nature exists on a continuum or in varying degrees. In addition to wilderness, there are, said Jacqui (Hobart), "all sorts of other types of nature," such as "artificial nature" (Colin/Perth), "altered nature" (Claire/Hobart), and "semi-natural" spaces (Paul/Hobart). Thus, Jo (Hobart) explained that, though urban parks populated by introduced species are "not what nature is" in Australia, she did not consider such environments unnatural, because "we, as part of nature, introduced it." The heavy-set mountain whose lower eastern flanks are studded with suburbs and whose western flanks lead out toward the Tasmanian Wilderness World Heritage Area was important for most Hobart participants in binding together city and wilderness. "One of the best things about living in Hobart," said Barbara, "is seeing Mount Wellington every day, and I make a point before I go into the office every morning just standing there and looking at the mountain for a minute, just to remind myself of the real world." Julian (Perth), who "avoid[s] the word nature because... it's so big as to be useless," offered an account in which the natural and the unnatu-ralm"meet, cross over, or are totally juxtaposed together in strange ways." The example he gave was of watching dolphins at play among the pylons in Perth's Swan River—a memorable experience that reflected his fasci-nation with wild nature in the city.

Collectively, what emerges from the interviews is an account of lived suburban space as artificial yet shot through with more-than-human reality. This lived space is saturated with technologies and techniques, though participants avoided these terms in describing their often highly deliberate practical engagements with the design, function, and mainte-nance of their houses and gardens. Home life emerged in a number of interviews as a skilled craft, one that maintained delicate connections with a living world. Descriptive accounts of encounters with the non-human agency of animals, plants, landscapes, skyscapes, and waterscapes were at the same time implicit accounts of careful and thoughtful technological practice. Consider how Greg (Hobart), now in his thirties, described — with enthusiasm delightful to witness—the suburban garden of his Melbourne childhood:

> And my Dad was a builder of things. He built a fishpond...and we had axolotls in the bottom in the mud and they lived there for quite a few years, and never saw them. We knew they were there... We had golden bell frogs... in an aquarium...they kept on climbing up the side and getting out. (*laughs*) Yeah, they just squeezed through the tiniest space, they were amazing. We had five of them. Oh, they are just wonderful little frogs, but they didn't stay little, they got quite big and the older they got the more brown they got. Dad built the pond with upturned bricks with the hollow underneath...When they didn't want to be seen...they'd sit in there in the hollow, and when it was sunny they would come out and they'd sit all round on the edge of the brick on top because it was nice and warm...And we even had white-faced heron visit our yard. So, it was pretty special...It had a very big influence on my life... We had the front yard full of natures.

Memorable encounters with frogs (and fathers); the smell of the sea (Kylie, for one, has to live within "sniffing distance" of the coast); the vast-ness of the sky (Chris, who lives in a mountainside suburb of Hobart, "quite often" thinks, "Wow, look at all that sky"): these were orienting points of contact with a more-than-human suburban horizon. These encounters were, however, also presented by participants as personal experiences bearing little relationship to their politics of nature, which revolved around a binary understanding of artifice and nature. In the final section I reflect on possibilities for an environmental politics of Australian cities capable of learning from and doing justice to the complex materiality of everyday life in suburbs and other hybrid, technonatural spaces.

Toward a Technonatural Politics

The foregoing section suggested the presence of a disjuncture between lived encounters with suburban nature and the politics of nature in Australian environmentalism. This disjuncture enables environmentalist discourse to carry criticism of the lives of many Australians without requiring environmentalists to reflect on their own lives. Participants' criticism of consumerism, individualism, and technological alienation slipped past accounts of the everyday practices that afforded them a sense of space, peace, privacy, and autonomy in the city, practices vital to their sense of orientation within more-than-human horizons.

The disjuncture between political categories and everyday experience was related to a global imaginary that saw the planet under attack from a human pestilence (Davison 2008b). The opportunity for this abstract imaginary to develop into a misanthropic politics was avoided in the interviews by the way suburbs stood in place for actual people during criticisms of human greed and selfishness. Reified as invading enemies of nature, suburbs were effectively erased of legitimate human desires and fears. The environmental behaviours disparaged as suburban were represented as the thoughtless and crude products of suburban environments rather than as the sources of those environments. In their own home environments, the participants were engaged in thoughtful and careful deliberation about the environmental implications of their domestic actions, in everything from decisions about what to eat to observations of local wildlife. Engaged in conscious practical relations with the world around them, they were inclined to present their home environments as fortunate aberrations from their wider narrative of suburban decline. In expressing their sense of good fortune to be living non-suburban lives in suburbs, few participants reflected about the role played by their practical deliberations in creating and maintaining a sense of contact with the world of nature in the city. Indeed, when discussing their practices, it was more common for them to intimate guilt for participating in activities they identified with suburban alienation, such as reliance on private automobiles or gardening with non-native species. For some, this guilty conscience took the form of an acknowledgment of failure to act purely, according to prescriptions derived from scientific descriptions of nature.

I take the view that significant benefits will flow to Australian environmentalism from efforts to resist the deterministic accounts of suburban life that presently obscure the co-production of human and more-than-human complexity in suburbs. This resistance requires nothing less than a re-evaluation of the concepts of nature and technology that have oriented postwar environmental movements. Much of the groundwork for this resistance is in place. The recent theoretical interest in non-dualistic accounts of modern materiality—an interest that underpins the present book—suggests strategies for articulating the co-production of nature and technology. Just as important, there are encouraging signs of a practical interest in rediscovering and remaking nature in spaces long regarded as lost to technology, such as the city. In what follows I briefly consider how this nascent technonatural politics opens up opportunities for Australian environmentalism to engage more effectively with Australia's suburban worlds.

I have argued that suburbs have long embodied yearning for nature and disquiet about technology even while helping sustain the modern project of technological progress. The suburb has confounded modern discourse, producing disdain and confusion more often than insight. It is not just environmentalists who have struggled to make sense of suburban spaces that are, ironically, so familiar in everyday life. Suburbs are one of many examples of the coupling of theoretical certainties and practical ambivalence in modernity. This history of lived ambivalence has been hidden by the failure of modern theory to account for hybrid materiality. As Latour (1993) has argued, the practical application of pure reason has been anything but pure. The scientific epistemology of dualistic categories was from the beginning joined to the production of hybrid entities and half-worlds that elude such categories. The worlds built according to the blueprints of pure reason have become increasingly unintelligible to it. In this context, the anxiety of environmentalists that they are unable to practise what they preach belongs to a wider dynamic of modern experience. Environmentalists' efforts to defend nature from technology or to rely on technology to save nature inevitably produce ironical results in practice, for the ideas of technology and nature on which they rely do not belong to the realm of practice.

I take the ironical nature of Australian environmentalist critique of suburban lives not as a marker of failure, but as a prompt for rooting new ideas of nature and technology in the ground of practice. Such conceptual possibilities are in fact already being embodied in Australian cities with greater confidence. Consider, for example, recent signs of the emergence of a popular urban ecological imaginary. A notable feature of the growth of scientific interest in the ecology of Australian cities over the past decade (Davison and Ridder 2006) has been the eagerness of residents to have confirmed what they have known in everyday experience — namely, that wildness is not absent from Australian cities and that non-humans have been paying little heed to scientific rules of engagement with humans. The conclusion of ecologists studying Australian cities is that they are "far more significant, ecologically, than most of us think" and that "nature is seldom as natural as we think" (Low 2002, 57, 106). Indicating that these lessons may already be established in everyday life, the response to the Australian Broadcasting Commission's invitation to its public to imagine what might be learned "if we all really looked at our own backyards — 20 million pairs of eyes across Australia looking at what's living there" (ABC n.d.) exceeded all expectations of participation. The 27,000 responses to this 2004 survey provided story after story of nature behaving unnaturally in and around Australian homes. They also led to the eye-catching finding that, despite the disapproval of nature-conservation professionals, "40 to 60 per cent of people in any street anywhere are actively feeding wildlife" (Murphy 2005, 9), giving weight to the suggestion that the "boundary between pet and wild creature" is becoming blurred (Low 2002, 121).

Consider two more ways that the boundary between nature and artifice is being blurred in practice in suburban Australia. First, the demarcation between gardening and ecology is being eroded by the many books, magazines, and television programs that now bring ecological themes to bear on Australia's still most popular recreation. Early destabilization of this boundary was achieved by the Permaculture and community-garden movements that emerged in the 1970s in Australian cities (Gaynor 2006; Timms 2006). More recently this erosion has gained impetus from one of the most significant, if still undocumented, environmental movements to appear in Australia in recent times. The urban landcare movement, comprising tens of thousands of volunteers working in autonomous local groups across metropolitan Australia, has taken shape over the past fifteen years (Davison and Ridder 2006). These groups seek to apply in practice scientific ideas of ecological restoration that draw sharp divisions between native and non-native species and pre-colonial and colonial times. Yet these groups also implicitly draw practical inspiration from the suburban gardener's familiarity with weeding. Focused on re-creating native environments, these groups are paradoxically placing human agency deep within this new nature, in the process training suburban residents in the practical art of making ecosystems.

Second, the boundary between domestic life and technological management is being blurred by pressure on suburban residents to practise domestic water and energy conservation. Directed by government incentives and regulations and spurred on by media reports that the nation is confronting potentially catastrophic climate change, householders in the early years of the twenty-first century have become central actors in efforts to decrease the resource appetite and environmental impacts of Australia's cities. This entry of a Promethean emphasis on eco-efficiency into the domestic sphere again confounds environmentalists' efforts to keep separate their agendas for nature and for technology. There is little recognition that recent emphasis on careful domestic stewardship of resources and wastes harks back to the suburban peasantry that once defined Australian cities. There is, however, growing recognition that Australian's suburban form offers ample opportunities for decentralized systems of water harvesting, solar energy capture, and localized food production, among other practices of domestic resource management (Troy 2003).

Environmentalists have pioneered many of the changes outlined above. Yet the politics of nature continue to be characterized in Australia by polarized and adversarial conflict between environmentalism and the social majority, many of whom — despite now sharing many concerns with environmentalists — continue to regard environmentalists with suspicion. I have argued that the failure to develop a technonatural politics of suburban life has seen environmentalists contribute to a paradoxical history of anti-suburban in a suburban nation. This failure has greatly limited the capacity of environmentalists to bring Australian society along with them in their quest for sustainability.

However, in reporting on interviews with environmentalists living in cities, I observed that, contrary to their bleak view of suburban life, many environmentalists are engaged in supple and creative practical negotiations with everyday suburban realities. My purpose has not been to dismiss the substance of environmentalist critiques of suburban life, but to catch sight of political strategies for contesting the future of Australia's suburbs that reflect their tangled, technonatural materiality. Such strategies begin with recognition of how much the lives of environmentalists share in common with those who do not share their politics. Relinquishing claims to purity, and thus freeing themselves from the spectre of impurity that shadows their present politics, these strategies promise environmentalists less guilt and greater capacity to see practical, sustainable possibilities for care of nature and for careful technology in Australia.

Acknowledgments

My thanks to the interviewees for their time and goodwill; to Zanni Waldstein and Merrin Ploughman for transcription; to Murdoch University for support in Perth; and to the Australian Research Council (Grant DP0344074).

Works Cited

ABC. n.d. WildWatch Australia. Australian Broadcasting Commission, Natural History Unit. http://www.abc.net.au/tv/wildwatch/archive/about.htm.

ABS. 2004. "Environmental Issues: People's Views and Practices." 4602.0. Canberra: Australian Bureau of Statistics.

Blühdorn, I., and I. Welsh. 2007. "Eco-Politics Beyond the Paradigm of Sustainability: A Conceptual Framework and Research Agenda." *Environmental Politics* 16, no. 2: 185–205.

Boyd, R. 1952. *Australia's Home: Its Origins, Builders, and Occupiers.* Melbourne: Melbourne University Press.

Braun, B., and N. Castree, eds. 1998. *Remaking Reality: Nature at the Millennium.* London and New York: Routledge.

Butlin, N.G. 1964. *Investment in Australian Economic Development 1861–1900.* Cambridge: Cambridge University Press.

Chaloupka, W. 2008. "The Environmentalist: 'What Is to Be Done?'" *Environmental Politics* 17, no. 2: 237–53.

Cronon, W., ed. 1996. *Uncommon Ground: Rethinking the Human Place in Nature.* New York: W.W. Norton.

Davidoff, L., and C. Hall. 1987. *Family Fortunes: Men and Women of the English Middle Class, 1780–1850.* London: Hutchinson.

Davison, A. 2008a. "Ruling the Future? Heretical Reflections on Technology and Other Secular Religions of Sustainability." *Worldviews* 12 (in press).

———. 2008b. "The Trouble with Nature: Ambivalence in the Lives of Urban Australian Environmentalists." *Geoforum* 39, no. 3: 1284–95.

———. 2006. "Stuck in a Cul-de-Sac: Suburban History and Urban Sustainability in Australia." *Urban Policy and Research* 24, no. 2: 201–16.

———. 2001. *Technology and the Contested Meanings of Sustainability*. Albany, NY: SUNY Press.

Davison, A., and B. Ridder. 2006. "Turbulent Times for Urban Nature: Conserving and Reinventing Nature in Australian Cities." *Australian Zoologist* 33, no. 3: 306–14.

Davison, G. 1995. "Australia — The First Suburban Nation." *Journal of Urban History* 22, no. 1: 40–74.

Desfor, G., and R. Keil. 2004. *Nature and the City: Making Environmental Policy in Toronto and Los Angeles*. Tuscon: University of Arizona Press.

Doyle, T. 2000. *Green Power: The Environment Movement in Australia*. Sydney: UNSW Press.

Devlin-Glass, F. 1994. "'Mythologising Spaces': Representing the City in Australian Literature." In *Suburban Dreaming: An Interdisciplinary Approach to Australian Cities*, ed. L.C. Johnson. Melbourne: Deakin University Press.

Fishman, R. 1987. *Bourgeois Utopias: The Rise and Fall of Suburbia*. New York: Basic.

Folke, C. 2006. "Resilience: The Emergence of a Perspective for Social–Ecological Systems Analysis." *Global Environmental Change* 16: 253–67.

Gascoigne, J. 2002. *The Enlightenment and the Origins of European Australia*. Cambridge: Cambridge University Press.

Gaynor, A. 2006. *Harvest of the Suburbs: An Environmental History of Growing Food in Australian Cities*. Perth: University of Western Australia Press.

Gilbert, A. 1988. "The Roots of Anti-Suburbanism in Australia." In *Australian Cultural History*, ed. S.L. Goldberg and F.B. Smith. Cambridge: Cambridge University Press. 33–49.

Gleeson, B. 2006. *Australian Heartlands: Making Space for Hope in the Suburbs*. Sydney: Allen and Unwin.

Hawken, P., A.B. Lovins, and H.L. Lovins. 1999. *Natural Capitalism: The Next Industrial Revolution*. London: Earthscan.

Hay, P. 2002. *Main Currents in Western Environmental Thought*. Sydney: UNSW Press.

Heynen, N., M. Kaika, and E. Swyngedouw, eds. 2006. *In the Nature of Cities: Urban Political Ecology and the Politics of Urban Metabolism*. London and New York: Routledge.

Hogan, T. 2003. "'Nature Strip': Australian Suburbia and the Enculturation of Nature." *Thesis Eleven* 74, no. 3: 54–74.

Horkheimer, M., and T. Adorno. 1972. *Dialectic of Enlightenment*. New York: Continuum.

Hutton, D., and L. Connors. 1999. *A History of the Australian Environment Movement*. Cambridge: Cambridge University Press.

Isenberg, A.C., ed. 2006. *The Nature of Cities*. Rochester, NY: University of Rochester Press.

Latour, B. 2004. *Politics of Nature: How to Bring the Sciences into Democracy*. Cambridge, MA: Harvard University Press.

———. 1993. *We Have Never Been Modern*. Cambridge, MA: Harvard University Press.

Light, A. 2001. "The Urban Blind Spot in Environmental Ethics. *Environmental Politics* 10, no. 1: 7–35.

Lines, W. 2006. *Patriots: Defending Australia's Natural Heritage*. St. Lucia: Queensland University Press.

Low, T. 2002. *The New Nature: Winners and Losers in Wild Australia*. Melbourne: Penguin.

McKibben, B. 2003. *Enough: Staying Human in an Engineered Age*. New York: Times Books.

McKinney, M. 2008. "Effects of Urbanization on Species Richness: A Review of Plants and Animals." *Urban Ecosystems* 11: 161–76.

Miller, J.R., and R.J. Hobbs. 2002. "Conservation Where People Live and Work." *Conservation Biology* 16: 330–37.

Milton, K. 2002. *Loving Nature: Towards an Ecology of Emotion*. London and New York: Routledge.

Morgan-Poll. 2000. "Australians Find It Easy Being Green." Finding No. 3309. http://www.roymorgan.com/news/polls/2000/3309.

Mullins, P. 1981. "Theoretical Perspectives on Australian Urbanisation: I. Material Components in the Reproduction of Australian Labour Power. *Australian and New Zealand Journal of Sociology* 17, no. 1: 65–76.

Murphy, K. 2005. "Wild about the City." *Melbourne Times,* June 22.

Newman, P., and J. Kenworthy. 1999. *Sustainability and Cities: Overcoming Automobile Dependence*. Washington, DC: Island Press.

Pakulski, J., and B. Tranter. 2004. Environmentalism and Social Differentiation. *Journal of Sociology* 40, no. 3: 221–35.

Parliament of Australia. 2005. *Sustainable Cities: House of Representatives Standing Committee on Environment and Heritage*. Canberra: Commonwealth of Australia.

Schlosberg, D., and E. Bomberg. 2008. "Perspectives on American Environmentalism." *Environmental Politics* 17, no. 2: 187–99.

Sessions, G. 2006. "Wildness, Cyborgs, and Our Ecological Future: Reassessing the Deep Ecology Movement." *The Trumpeter* 22, no. 2: 121–82.

Shellenberger, M., and T. Nordhaus. 2005. "The Death of Environmentalism: Global Warming Politics in a Post-Environmental World." *Grist* [online journal]. http://www.grist.org/news/maindish/2005/01/13/doe-reprint.

Taylor, C. 1989. *Sources of the Self: The Making of the Modern Identity.* Cambridge, MA: Harvard University Press.

Timms, P. 2006. *Australia's Quarter Acre: The Story of the Ordinary Suburban Garden.* Melbourne: Miegunyah.

Troy, P. 2003. "Saving Our Cities with Suburbs." In *Griffith Review: Dreams of Land,* ed. Julianne Schultz. Brisbane: Griffith University. 115–27.

———. 2000. "Suburbs of Acquiescence, Suburbs of Protest." *Housing Studies* 15, no. 5: 717–38.

Weber, A.F. 1969[1899]. *The Growth of Cities in the Nineteenth Century: A Study in Statistics.* New York: Greenwood.

Williams, R. 1973. *The Country and the City.* London: Chatto and Windus.

Part Three

Technonatural Present-Futures

Chapter Eight

The Property Boundaries/Boundary Properties in Technonature Studies: "Inventing the Future"
Timothy W. Luke

This chapter investigates the boundary properties of technonature by probing the property boundaries of rapid "citification" and "deruralization" that is being created by transnational capitalism. It begins by reconsidering how global climate change is tied to the greenhouse gases generated by humans in their everyday lives; it considers whether these greenhouse gases have actually created a gaseous greenhouse in which technonature can be directly apprehended and assessed. In suggesting that this development is becoming more real with each passing year, the analysis then argues that the spatial practices of transnational capitalist exchange are the accidental normality of our environment, which reveals its real breadth and depth only intermittently as "normal accidents."

Instead of anticipating and avoiding these events, the hybridizing co-evolution of humans and machines accelerates their spread, incidence, and effects by organizing global exchange around continuous "innovation" apparently in itself and for its own sake. This innovation, however, is also organized to ensure that its accidental normality advances the tighter governance of people and things through the contact of things and people in the right disposition of conduct as it "invents the future." With Nature so closely ordered, and Society so fully contained with the hybridities of technonature, it becomes useful to appraise the contours in these multiple interconnections as the new spatial practices of an "urbanatura" to the degree that the accidental normality of greenhouse-gassing global capitalism envelops humans, non-humans, and hybrids in technonatu-ralized systems and structures.

A crucial beginning is to examine how the disorder of contemporary urban industrial production and consumption routinely creates more and more noxious by-products. These wastes lodge in the Earth's soil, atmosphere, water, and living inhabitants, both human and non-human, and travel, both locally and globally, out along unsustainable worldwide webs of economic exchange until they create new material environments. If one assumes that atmospheric change today, which has been socially constructed as "global warming" in studies of the "greenhouse effect" (Long 2004), is in fact a material reality—albeit of contingent duration, scope, and effect—then one is setting forth new coordinates for a system-atic critique. Some see a crisis looming for environmentalist theory and practice if it fails to face these realities about global warming (indeed, some, such as Shellenberger and Nordhaus [2004] see the "death of envi-ronmentalism"), but their analysis is flawed (Luke 2005). The questions here are much deeper, more interesting, and more important.

The social construction and creation of "property boundaries" for technonature, in and around urban-industrial formations where noxious products and by-products are produced and consumed, is the initial focus of this chapter. If one takes the "greenhouse effect" seriously as a materially evolving reality, then all "effects" in/of/from "the greenhouse" allow us a chance to reassess our conventionally stabilized Society/Nature concepts in a more critical manner. While rarely addressed systematically by most "green-grounded" environmental criticism, these developments must be examined carefully through more radical approaches to such "grey-zoned" environs in studies of "cyborganization," "socio-nature" or "techno-nature." Whether they examine technoscience operations, natu-ral disasters, or socio-systemic collapses, some studies (Castree 1995; Luke 1999a, 1999b, 2001; Swyngedouw 2004) scan the "property bound-aries" of urban space as they are stabilized in ordinary policy terms such as urbanization, land use, environment, river basins, industrialization, economic growth, sprawl, or natural resources. Once scrutinized more closely, the unstable, unconventional, and undetected properties of mul-tiple industrial hybridities do emerge out of foggy phenomena, including the "greenhouse effect."

Technonature: Global Climate Change

To explore technonature, one must dismantle today's liberal-democratic notions of humanist agency and naturalized structure in search of hybridizing influences at play in actually existing democratic consumer capitalism. For years, Donna Haraway has been tracking the disruptive patterns of technonature with her vision of cyborgs, which come together at sites and systems characterized by

> continued erosion of the welfare state; decentralizations with increased surveillance and control; citizenship by telematics; imperialism and political power broadly in the form of information rich/information poor differentiation; increased high-tech militarization increasingly opposed by many social groups; reduction of civil service jobs as a result of the growing capital intensification of office work, with implications for occupational mobility for women for colour; growing privatization of material and ideological life and culture; close integration of privatization and militarization, the high-tech forms of bourgeois capitalist personal and public life; invisibility of different social groups to each other, linked to psychological mechanisms of belief in abstract enemies. (Haraway 1991, 171)

Haraway's notion of these cyborganization processes clearly calls attention to the unfixed boundary properties of nature/culture, city/country, human/non-human, where capital and power continuously define property boundaries. Her initial efforts at disclosing these cyborganic dimensions of life were written self-consciously as jeremiads (Bartsch, DiPalma, and Sells 2001, 127–64), because this discursive genre purposely is dialectical. Her cyborg concepts fuse the promise of latent possibilities with the problems of extant experiences to assail the disappointments of the present in discursive exercises. To better understand technonature, as this collection suggests, more existing social relations of cyborganization need to be traced to some of their cyborganic sources.

When examining greenhouse effects, for example, one must go back to the future to see technonature evolving. Many accept the historical convention that the Industrial Revolution began in the eighteenth century as steam engines and growing cities led to tremendous increases in the consumption of coal, wood, and, later, petroleum-based fuels to generate the energy needed for modern industrial life. Geological, botanical, and oceanographic evidence from that time forward reveals increasing levels of carbon dioxide and other greenhouse gases concentrating in the atmosphere. To document this shift, however, Nature itself must be reified, and then reduced by way of operational models of mensuration as a materialized technonatural formation. A reimagining of the Earth as a fixed structure of infrastructural systems and spaces producing ecological services was carried out as early as 1824 in a scientific study by Jean-Baptiste-Joseph Fourier. His study, "General Remarks on the Temperature of the Terrestrial Globe and Planetary Spaces," recast the biophysics of atmospheric chemistry, solar radiation, and terrestrial temperature as operating interdependently, essentially like a giant glass dome, to generate the warmth needed to sustain the biosphere and all its human and non-human inhabitants (Long 2004, 61–63).

Fourier developed his scientific insights during the last decades of the "Little Ice Age" of 1300–1850. His fascination with how global warmth is maintained is not surprising, but he was essentially imagining a re-engineering of natural environments as technonatural systems for the advancement of humanity on Earth. Other scientists in Sweden and Great Britain maintained this technonatural take on global warming in studies that tied the Earth's geological cycles of ice ages and warmer interglacial eras to variations in the levels of carbon dioxide in the planet's atmosphere. By the 1860s this research had recognized that water vapour and ozone help absorb and retain heat; however, scientists in Great Britain—then the "workshop of the world"—argued that increasing carbon dioxide emissions from anthropogenic sources could actually enhance this greenhouse effect, sustain global warming, and improve weather conditions for humanity in a vast technonatural design.

As the Industrial Revolution unfolded, increases in fossil-fuel use by humans created a virtual greenhouse, so that average planetary temperatures are now rising. Some scientists regarded these trends as positive, believing that the boundary properties of this technonature would improve agriculture, postpone the world's periodic reglaciation, and maintain better living conditions than mere Nature itself had done (ibid.). With industrialization, the Earth was "greenhouse gassed"; industrial effluents virtually remade the planet as a "gas greenhouse"—that is, a technonature.

Implicitly, then, the boundary properties of nature/society, city/country, urban/rural, were being contested soon after industrialization began.

The property boundaries of cities were exceeded and sublated by their noxious by-products as well as by their beneficent products. As urban fossil-fuel wastes accumulated in the atmosphere, oceans, and soil, the Earth itself was recast in the scientific imagination as artifice, architecture, or artifact, in that greenhouse gases were indeed concentrating in the atmosphere as the vaporous gridworks of a terrestrial greenhouse. The technical and the natural were in this way (con)fused, and technonature arose from the smoke and ashes.

Strangely, the property boundaries of nature/society are often frozen in time and defended with greater vigour in order to somehow forestall the acceptance of technonature by ignoring it. For example, a recent, widely read manifesto, "The Death of Environmentalism" (Shellenberger and Nordhaus, 2004), attempts to focus today's widespread uneasiness among the public by slamming the social philosophies and political practices of mainstream environmental activists in the United States. Shellenberger and Nordhaus (2004, 6) acknowledge the past achievements of those activists, but they also assert that "modern environmentalism is no longer capable of dealing with the world's most serious ecological crisis." Believing that this crisis can be reduced to what they glibly represent as the new threat of "global climate change," and that this threat can be resolved only by stopping anthropogenic "global warming," they argue that traditional environmentalist movements have neither recognized nor met the challenge. These groups have failed because "environmentalism has become a special interest" that has sold out to a liberal politics in which all activist groups become "special interests" (ibid., 28) for preserving an unsullied Nature.

Even as they pillory environmentalism for struggling to preserve the wilds by capitulating to private capital, embracing state regulation, or settling for resource conservation via technical innovation in a truce with technonature, Shellenberger and Nordhaus claim that the only real escape left from global warming—which they regard as a severe "ecological surprise" (King 1995)—is to copy the "successes" of right-wing political activism since the 1970s by capitulating to the experts and owners behind transnational capitalism. This move would entail riding a new "third wave of environmentalism [that] will be framed around investment" in new public/private partnerships "like those America made in the railroads, the highways, the electronic industry, and the Internet" (Shellenberger and Nordhaus 2004, 28). One can sympathize with their disappointment over the conventional eco-managerialism favoured by the National Environmental Protection Acts (NEPA) since the 1970s to preserve Nature; that said, their preferred solutions for answering global warming are based on complex corporate deals — "public–private investments" — of the sort that underpinned the growth of America's railways, interstate highways, and electronics industry. But consider: if these huge technological systems are what spew greenhouse gases into the atmosphere, how can they possibly provide ideal solutions for curbing global climate change? Either way, a pristine Nature cannot be preserved, and all of these new technological formations will continue advancing the evolution of technonature.

As various clusters of technologies, points of capital investment, and sites for reshaping the materiality of some industrial ecologies within Nature's biophysical systems, technonature exists. Hoffman (2003) maintains that in trying to revive "environmental protection" after the "death of environmentalism," Shellenberger's and Nordhaus's lament clearly fails to deal "not just with material considerations of the physical sciences but also with considerations of what motivates individual and organizational action found within the social sciences" (2003, 77).

Indeed, Shellenberger and Nordhaus show that in many American policy-implementation practices, there is an unclear understanding of how deeply private interests have burrowed into larger public projects in the economy and environment (Fischer 1990). Such miscomprehension is traceable to world views that ignore technonature by keeping Nature and Society active and separate. The liberal dictates of private property and the faith placed in technological expertise pit private interests and technical accreditation in "Society" against truly collective public concerns in "Nature" when, in fact, all is technonature. The challenge today is not how "to get back on the offensive" (Shellenberger and Nordhaus 2004, 29–31) by doing private conservative activists one better. Rather, it is to develop a truly *public* ecology with new organizations, institutions, and ideas whose material articulation can balance the insights of scientific experts, the concerns of those who hold private property, and worries about social inequity, with the need for ecological sustainability to support human and non-human life in the hybridized technonature of twenty-first-century life.

Light and Shippen contend that "the specific importance of environmental public goods" (2003, 234) makes it imperative to lay deep materialist foundations for protecting the community goods, collective needs, and public services of stable, healthy industrial/natural metabolisms for all human and non-human life. Today, however, only a technonaturally minded ecology can interpret effectively all the new strategies linking state, market, and civil-society approaches; only in this way can the most destructive urban-industrial ecologies of technonature be rethought and restructured. Otherwise, environmentalism will stay snared in the hidden and shifting agendas of neoclassical economics (the "private") — agendas that have rooted themselves throughout the world over the past two decades "in a variety of many governments' approaches to a variety of economic and policy matters based on this dominant economic view" (ibid., 232) — all the while continuing to occlude technonature.

The material basis of a fully identified technonature, then, can be tracked down in the everyday logistical complementarities, metabolic exchanges, and ecological roles materially made between co-evolving human and non-human communities (Virilio 1995, 2000). Such relations are spun forth from individuals and groups brought together in global ecologies. As cohabiting populations, co-operative producers, collaborative consumers, and correspondent groups, human beings openly form "publics" of property makers, owners, and users as well as shared "ecologies" of material creators, energy users, information exchangers, and waste generators. The spatialized practices of owned property, property owners, and proper ownership, in turn, compound the forces of material supply and demand to create technonature.

Materialities then set specific boundary properties for subjectivities-in-action as their inputs, throughputs, and outputs (both intended and unintended) cascade through the world's machinic webs of waste generation, information exchange, energy use, and material creation, which together set property boundaries of objectivities-for-action. While many would drape such webs with opaque veils of private-property ownership or special technified expertise, essentially they are open material spaces that define built spatiality, temporality, and activity for technonature. This "urbanatura" of technoculture/technonature is always already open and should not be closed. In many respects technonature can be defined through public ecologies. Such environmental formations are where the activities connected with sheer survival are permitted — yet also pushed forth — to appear in public, as public, and from the public via new institutions of public ecological co-operation (see Arendt 1998, 46).

Defining public ecology depends on recognizing how thoroughly the materialization of urbanatura marks a structural transformation of the planet's biosphere. Defending this public ecology — as it is extruded bio-culturally, sociospatially, and technoscientifically in the transformed structures of the biosphere/technosphere — requires a new sense of how the cultural, economic, and social forces infiltrating these built environments have fused the historical with the natural (Luke 1999b). These links are complex, but their recognition, articulation, and reoperationalization are imperative, first to reimagine and second to protect the biosphere as a continuously transformed public sphere (see King 1995: Luke 1999; Hoffman 2003).

Coping with higher carbon-dioxide levels in the atmosphere, for example, is now a general condition linking humans and non-humans. That is, greenhouse gases are a cyborganic quality of technonatural life under industrial capitalism. Haraway argues that a cyborg world must become centred on "lived social and bodily realities in which people are not afraid of the joint kinship with animals and machines, not afraid of permanently partial identities and contradictory standpoints" (1991, 154). Greenhouse gases, imperfectly but relentlessly, are forging these partial identities and contradictory qualities; all the while, living organisms and terrestrial features are warming up together as technified nature. Hydrocarbon wastes are turning modernizing global markets into a local earthly materiality; technonature is product and producer, act and agency, site and situation.

With technonature, the market mediations of buying and selling at billions of episodic points for exchange-driven valorization are nothing but permanently partial identities and contradictory standpoints in which both social and bodily realities also reveal technonaturalized materialities. As we track back cyborganic spaces of collective and individual hybridity, Haraway is among the first to wonder why "our machines are disturbingly lively, and we ourselves frighteningly inert" (ibid., 152) as technonature evolves. Can property boundaries adduce the boundary properties for the "human" that actually are "non-human" when technonature proliferates as "cyborganic" matter? Spatiality itself perhaps holds answers to such questions.

Accident Normality as Spatiality

Henri Lefebvre (1991, 38) contends that the analysts of economies and societies must examine "spatial practice," because it "secretes that society's space; it propounds and presupposes it, in a dialectical interaction." In today's neocapitalist order, spatial practice "embodies a close association, within perceived space, between daily reality (daily routine) and urban reality (the routes and networks that which link up the spaces set aside for work, 'private' life and leisure" (ibid.). Arguably, with technonature/technoculture, these materialities of spatial practice are foundational. With technonature as spatiality, then, one can begin to explore urbanatura as an interplay of practice, thought, and activity "which exists within the triad of the perceived, the conceived, and the lived" (ibid., 39).

In a very real material sense, the forms and flows of the human and non-human hybridity of urbanatura concretize Lefebvre's visions of the "urban revolution" (2003[1970])—visions that are unfolding all over the planet as public ecologies. Its spaces are entirely built environments, tied into complex layers of technological systems whose logistical skeins are knit into unbuilt natural systems and sites for the production, consumption, circulation, and accumulation of commodities. For Lefebvre these embedded cycles of commercial existence are force fields for "the urban," infiltrating on a global scale what once was seen on a local scale as the divided realms of the social and natural, countrified and citified, peasant and proletarian, which all have their own differential, disjunctive, and discontinuous complexities. In past attempts to gauge such spatiality, an analytical blindness induced by sutures in time or twists in history often occluded what was at hand. Lefebvre maintains that "the urban (urban space, urban landscape) remains unseen" (ibid., 29) because human cognition and perception are trained to be incapable of sighting it.

Public ecology is perhaps one of the most solid material opportunities to overcome such blindness. It enables us to cut new sets of spatial optics to detect the disruptive snarls in the worldwide webs of natural and social exchanges (see Dodds with Middleton 2000)—snarls that create destructive outcomes such as inequity and inefficiency in technonature. In many ways the citification and deruralization of urbanatura are material concretions of public ecologies past, articulations for public ecologies present, and projections about public ecologies future. Some of these linkages now appear as the ill effects of global warming, but they hardly constitute Shellenberger and Nordhaus's post-environmental world. In fact, the enduring material elements in technonaturalized urbanatura link the co-evolution of human beings and non-human things in many "social" and "natural" ecologies between, behind, and beneath commodification, which must be made more open, less closed, and therefore transparently public. Cities remain pivotal sites where everyday exchanges between built and unbuilt environments occur, but these transformations also transpire in urbanatura's deruralizing metabolic reactions in the countryside.

Urbanatura, then, knits together what once were quite discrete natural and social elements into the compound prefigurations of a planetwide community, which Lefebvre in 1970 termed "the urban." What seemed only like "a horizon, an illuminating virtuality" (Lefebvre 2003[1970], 17) a generation ago is now fully recognizable in today's deruralizing and overurbanized ways of life. While "the urban" is perhaps somewhat abstract, its fields of force remain the principal focus of human ecological action — in cities and countrysides alike. Consequently, the material impulse of capitalist globalism to remake the world in commodified forms through markets also is being countered by dense webs of social movements resisting the urban revolution's inequitable and inefficient reconstitution of the world's environments and their inhabitants as urbanatura.

One place to begin mapping urbanatura more directly as technonaturalized spatial practice is Charles Perrow's *Normal Accidents: Living with High-Risk Technologies* (1984). Perrow makes some intriguing observations in that book about how new, unfixed technologies such as electricity, chemicals, space travel, and air transport inherently involve high-risk applications in their spreading use; but he also finds that most of the risk can be placed at the doorstep of their applied systemicity. That is, as technonatural processes unfold, "we create systems — organizations, and organizations of organizations — that increase risk for the operators, passengers, innocent bystanders, and for future generations" (1984, 3).

The ordinary material operations of risk-ridden enterprises carry with them a "catastrophic potential," yet "every year there are more such systems" (ibid., 3). Events that Perrow labels as "normal accidents" occur because they are simply so many innumerable conjunctures of contingency needed by each system in order for the system of systems to work. Inevitably, there are moments when such work fails — when it exceeds, lags behind, or otherwise deviates from ordinary parameters just enough to transmogrify acceptable ordinary actualities into unacceptable catastrophic potentialities. The "normal accident" then occurs.

These events are arresting and awful in part because they reveal how much the raw secretions of technonatural space rest within a circumambient "accidental normality" of risk-ridden system building. Fields of order as well as disorder lie beside, beneath, or behind the many "normal accidents" that are latent within technonatural machinations. Systems get built. Their construction usually is fitful, irrational, unsystematic, and improvisational; even so, these provisionalities stabilize to the point of becoming "normal." An obvious example is the greenhouse gassing that has been accepted as a general background condition of industrialization since the 1820s. Still, the normality — for all of its apparent elegance, power, or order — remains as much accidental as it is intentional. Perrow argues that one can try to calculate the probabilities of accidents, manage their risks, and contain their damage. But this effort can only achieve so much: technonature is accidental and therefore always partly unpredictable. However, Perrow misses the key importance of technonature's accidental normality — that is, the degree to which its spatial materialities ordinarily continue every day except for a few contingent occasional failures.

Hence, to anticipate the unusual incidence of "normal accidents" in large, complex systems — as many risk-assessment exercises have done — is to also participate in the generation, indeed the naturalization, of this ordinary "accidental normality," which lies at the core of systematized large complexity. Technonature clearly becomes "the perceived, the conceived, and the lived" (Lefebvre 1991, 39) in the folds of technonatural spatiality. In some ways the intrinsic physical, operational, mechanical, chemical, or biological qualities of many technologies direct engineering toward particular paths of creation; this makes normal accidents possible. In attempts to live with these normal accidents, however, one cannot ignore how a vast accidental normality always awaits possible catastrophic eventuation from interstices of everyday life.

Living in societies of bureaucratically controlled consumption (Lefebvre 1984) on a transnational scale discloses that consumption is a normative cluster of conduct that directly enables modes of bureaucratic control and control by corporate, government, and technoscientific bureaucracies. To examine "everyday life in the modern world" is to realize how much "the modern world" is an imagined, embedded, and engineered community that normatively delimits, defines, and directs "everyday life" as the accidentally normalized forms of life found in technonature. Industrial products, industrial processes, and industrial production are systems of conducting conduct by managing fear, insecurity, and desire. Technonature is far from natural but is not yet wholly artificial. Rather, it is a manifold hybridity of engineered, embedded, and imagined spaces in which the quality, pace, substance, and opportunity of practices that define material and mental life derive from decisions (as well as non-decisions) from elsewhere (taken or not taken by unknown others) without much popular participation, deliberation, or even awareness. In other words, perceived, conceived, and lived space—both citified and derural-ized—unfolds as our accidental normality (Mumford 1961, 322–24).

"Nature," however, can still swirl about as "technonature" flows along. On this point, "certainly we continue to have," as Hardt and Negri argue, "crickets and thunderstorms...and we continue to understand our psyches as driven by natural instincts and passions; but we have no nature in the sense that these forces and phenomena are no longer understood as outside, that is, they are not seen as original and independent of the civil order" (2000, 187). The daily drift of materiel, system, and artifact, as these fold into technonature, is still where webs of governmentality truly gel, even as clouds and crickets fly around.

"Innovation" as Technonature

The emergence of cyborganization narratives during the space race of the 1950s and 1960s, followed by the fixation on those narratives as objects of debate as the Cold War wound down, underscores how society now worries far more about technonature by "inventing the future." As *Forbes* magazine asserted in its December 2002 celebration of "Human Innovation," these fixations on the future preoccupy more people than do apprehensions about sliding back into the past of Nature, which apparently is gone for good in "the postmodern condition" (Jameson 1991).

Playing off this construction/creation contradiction, do the processes of technonature become more transparent in everyday tests to find boundary properties in the manufacturing and marketing of things by people, for people, with people? To speak of the fusion of men and machines, the natural and the artificial, humans and animals, is to refer to the inter-operation of a person with a process, a precept or a practice in some machinic artifice, one that governs time, matter, intelligence, and energy. *Forbes,* a popular American business magazine, asserts that capitalism has outlived communism because "the market" performs these practices—easily, smoothly, and every day. *Forbes* celebrates its publishing achievements as "a capitalist tool" by constantly hyperhumanizing the practices of technological innovation by creating and constructing the hybridities of technonature. This practice raises questions. How much can a cyborg be an organism of/in/under governmentality, whose conduct is being conducted through some design for governance? Its designations of domination, assignations of authority, and consignments of control often are the boundary properties of interoperation where technonatural property boundaries emerge. Much of this conceptual concretion spins up from the froth of marketing; but at times of retrospective celebration, capital unconsciously concedes its co-evolution with labour in the consubstantiation of the commodity, which in turn (con)fuses human and non-human forces technonaturally. Likewise, technonature constitutes the accidental normality of innovation as invented past/present/future for all organisms in sites and spaces of coordinated systems for conducting conduct.

In this *Forbes* retrospective the (con)fusions of people and things, nations and markets, humans and machines, innovations and individuals, are counted off one by one and year by year over eighty-five years beginning in 1917 and ending in 2002. What is truly a highly accidental normality becomes recast rationally as innovative futurism. Change is presented as neither accidental nor normal. Instead it allegedly sweeps along as instrumental rationality on a quite purposeful and extraordinary march through time as the goods, services, technologies, and practices of markets. From sneakers in 1917, to business management in 1923, to wallboard in 1930, to xeroxgraphy in 1938, to Tupperware in 1947, to Levittown in 1951, to integrated circuits in 1959, to Telstar in 1962, to the Internet in 1969, to discount brokerages in 1973, to junk bonds in 1977, to LCDs in 1984, to Prozac in 1987, to the World Wide Web in 1991, to automated gene sequencing in 2000 (Armstrong et al. 2002, 124–210), the property boundaries of markets generate the boundary properties of technonature's spatial practices. What others see as "cyborganization" *Forbes* deems "innovation," which entails markets activating ideas to remix the conduct of conduct for people and things by (con)fusing humans with non-humans in new, technonaturalized ways.

While not casting these developments as the work of a machinic collective, "the greatest breakthroughs of the last four score and five years: are meant to decenter the Enlightenment conceits of heroic humans and great geniuses, of minds — that is why you won't find Bill Gates on it. It is not a roster of the deserving (Gandhi) or the powerful (Stalin) or the biggest empire builders (Kaiser)" (ibid., 124). Thus the technonatural interactions of men and machines, nature and artifice, people and animals, disclose "a collection of people, products, services, and companies that have changed our lives in a profound way" (ibid.).

From the alchemy of capitalist exchange, the "85 Innovations" are, ironically, some roots of that accidental normality that is conducting the conduct of humans (Post 2002, 123). *Forbes* notes that it began publishing in 1917, "the same year as the Bolshevik revolution — and has outlived that experiment by nearly a dozen years" (ibid.). The Third International tried to sublate borders in a world of nations in turmoil. In this regard, *Forbes* celebrates how the accidentally normalizing forces of markets now are prevailing over boundaries where the Bolsheviks failed. Innovations and futures "recognize no boundaries; great ideas migrate across borders, despite all attempts to stop them" (ibid.).

This otherwise unremarkable business text, then, provides a cache of clues about urbanatura in that it celebrates "two forces that are bringing the world closer together: innovation and capitalism" (ibid.). The effusive ties of cyborganizing in *Forbes* are mystified by capitalist exchange at the individual level. The Special Anniversary issue's contents page promises to review "the things that changed the way we live" and at the same time to disclose the identities and locations of "the people who will reinvent the future." Here, traditional ideologies of methodological individualism presume that individuals are methodically reinventing things that will bring about the future through processes and products. At the collective level, in turn, these individuals and their methodologies are trapped in "the world," which is for *Forbes* in 2002, as it was in 1917, "in turmoil" (ibid., 123). Today, however, "the new ties that bind nations are based not on coercion" but rather "upon enlightened and mutual self-interest, promising greater stability and prosperity" through the conductors of things, which achieve in markets the conduction of commodification.

Markets may well remain inscrutable, but it is their elusive spatial practices that count. Their tracks are to be found in the times and places where firms, such as General Electric, put "imagination at work [to] bring good things to life" as technonature. *Forbes*, then, is implicitly asserting that these machinic collectives are what change our technonatural lives in a profound way: "It is a history of lightbulbs that went off and changed the world. The transistor gave rise to a trillion-dollar piece of the economy and a potent deflationary force. The pill altered human behavior; the polio vaccine and protease inhibitors altered life spans; the discount bro-kerage changed our capital markets" (ibid., 124).

Each year, in turn, is a marker for new property boundaries, for reveal-ing the things that changed the ways we live; but this sense of time starkly evades the issue of whether or not boundary properties of innovations bring greater stability and prosperity or move the world closer and faster toward turmoil. This process, while normal, is accidental.

Possessions possess their possessors, but does this possession stretch beyond the control of commodities into a regimen of governance? Boun-dary properties blur here as property boundaries are fixed by labour, cap-ital, and the market. Owning things requires knowing things; but what is this knowledge? Who or what generates it, and how does power arise along with this knowing? The systems of technonature join together dead labour with living labourers; and guidance comes from the dead hands of the past, which may well pull the hearts and minds pulsing with these energies and intentions through the firms whose products possess their users all across urbanatura (Virilio 2000). Technonatural cyborganics are always already at work in the multiplicities of mechanism and the plural-ities of practice. *Forbes*'s efforts to document the "85 Innovations That Changed the Way We Live" imply that purposive genius always must somehow be at work, finding the ways, forcing the changes, creating some we-ness, and defining what all of these human/non-human (con)fusions mean together as lifeways, lifeworks, lifeworlds in urbanatura. Its Special Anniversary issue presumes simply to celebrate such "innovation," but this move mystifies how the ordinary processes of innovation are every-day technonatural activities spun up from the market's exchange of prod-ucts and practices (Virilio 1995).

Hardt and Negri (2000) tacitly concur with technonatural co-evolu-tion—albeit without giving it this name—because in today's globalized economy and society, "inventing the future" has gone "beyond a funda-mental threshold here when the multitude recognizes itself as a machinic, when it conceives of the possibility of a new use of machines and technol-ogy in which the proletariat is not subsumed as 'variable capital,' as an integral part of the production of capital, but is an autonomous agent of production" (2000, 405). They hide their hearts in the hedges of "hybrid-ity"; even so, let it be said that the collective *telos* and teleological collec-tivism of technonature lie at the roots of the multitude's existence. For Hardt and Negri, "the hybridization of human and machine is no longer a process that takes place only on the margins of society, rather, it is a fundamental episode at the center of the constitution of the multitude and its power" (ibid., 405).

Hardt and Negri obliquely admit that to raise questions about technonaturalized becoming, "one speaks of a collective means of the constitution of a new world, one is speaking of the connection between the power of life and its political organization. The political, the social, the economic, and the vital all dwell together" (ibid., 406). Socialists and communists once called for the proletariat to gain complete control over the means of production; today even *Forbes* recognizes that the postmodern moment amounts to the collective constituting itself in all the markets, machines, and materials that people and things generate together. Hence "the multitude not only uses machines to produce, but also becomes increasingly machinic itself, as the means of production are increasingly integrated into the minds and bodies of the multitude. In this context reappropriation means having free access to and control over knowledge, information, communication, and affects" (ibid., 407).

Cyborganizing Governmentality

Technonature is also biopolitics in action. "To govern," Foucault asserts, "means to govern things," and corporate globalism's emphasis on performativity has led most firms to concur that "government is the right disposition of things, arranged so as to lead to a convenient end" (Foucault 1991, 94, 93). Commodities, as complex social and technological artifacts, provide cyborganic ensembles of multiform tactics for corporations, state bureaucracies, or professional experts to rightly dispose of things, arranging their boundary properties to lead the marketplace to many convenient ends with clear property boundaries.

The hybrid beings evolving at the specificities of each commodity/ user, or commodity/maker, nexus are not simply inert, passive receptors. He/she/it is an active, volatile capacitor for every circuit of corporate exchange effects in technonaturalized cyborganization cycles. As company growth targets circulate through innovative nets of accidental normalization, goods and services in the marketplace constitute both individuality and collectivity for tomorrow around the prevailing norms of consumption today, because "individuals are vehicles of power, not its points of articulation" (Foucault 1980, 98). Commodities work as effective relays of corporate management only insofar as their generic capacities for market-mediated individualism become materially articulated in their specific effects on one, some, many subjects as well as collectivized as the universal affects within one, some, many objects. The range of cyborganic subjectivities, then, is concursively cast, in part, at the cash and commodity nexus, with the objects produced in part by globalization (Luke 2001).

Hence capital's — and even labour's — celebratory wonder over the invented future, which delights all as accidental normality, pays out for some in modernity's improved levels of nutrition, housing, transportation, health, information, wealth, and communication. To a large extent, however, these social relations are an implicit appreciation for a world that, as *Forbes* reminds us, is now possessed by agendas framed for the utility of AT&T, Citibank, Exxon Mobil, Johnson & Johnson, TimeWarner, and Xerox. How thinking happens, how deeds are done, how knowing unfolds, therefore, cannot be separated from how ownership and control both operate. By rethinking how people who interlace their lives with technologies have "anonymous histories," one finds many approaches to technonature in action (Giedion 1948).

The end users of corporate commodities are redesignated through their purchases of products to play the role of capital assets, causing "the ultimate realization of the private individual as a productive force. The system of needs must wring liberty and pleasure from him as so many functional elements of the reproduction of the system of production and the relations of power that sanction it" (Baudrillard 1981, 85). Corporate plans for global transformation gain life, liberty, and property through the buying decisions of individuals rather than the other way around — which is the myth that *Forbes* pushes (Luke 1999a). For transnational businesses, liberating personal "wants" and individual "needs," as they are supposedly felt by everyone anywhere, entails making more and more commodities hitherto inaccessible in many markets available to all who desire them. In this way nature is technified, technology is naturalized, and those making the markets keep on "inventing the future" in the swarms of sales (Kelly 1994).

Truly the most significant "new social movements" at work today, as the *Forbes* celebration of "85 Innovations" indicates, are the multiple mass movements of technonaturalizing agents and structures — both public and private — that guide consumers through urbanatura from new commodity-spaces to newer commodity-spaces and on to the newest commodity-spaces. Global lifeworlds are now in the aggregate more grey-zoned environments composed of corporate institutions within which open-ended experiments with new artifact-acts follow experiences with older artifact-acts. Globalization, in these conditions, brings a subjectivity of object-centredness. Everyone is what they buy, and everyone buys what they are, have been, and will be (Agger 1989). "In this end," as the vision of corporate producers and consumers of products proves, cyborganized peoples become "destined to a certain mode of living or dying, as a function of the true discourses which are the bearers of the specific effects of power" (Foucault 1980, 94). Spatially, these effects transpire within urbanatura; but at the same time, such events generate urbanatura's technonatural qualities. Even so, this normality ultimately is accidental, but so too are catastrophic accidents quite normal everyday events.

Technonature marks how private ecology has turned the world's built and unbuilt environments into a formation that is of a piece in "urbanatura," and not two wholly separable entities, as with "Society" and "Nature." As Lefebvre suggests, private ecology exploits those mental divides as it degrades the overall civic life of society. Privileged millions continue to benefit from the international misery of billions; Shellenberger and Nordhaus's anti–global warming politics (2004) cannot help us find a post-environmental world by simply endorsing other networks for special-interest politics. Indeed, their eclectic belief in doing what right-wing activists have done for decades — only now with a different progressive spin and a broader popular coalition — could well lead back to a resurrection through which the non-environmentalists all suffered during the Gilded Age, before the first conservation movements formed in the United States.

Technonature spatially gels as the material means for conducting everyday affairs through markets and states by enabling alternative modes for living to present themselves in the operational utilities of cyborganic urbanism and ruralism (Virilio 1997). City people are different from villagers because the machinic systems of cities give bigger, broader, and perhaps even better logistical articulations to the common formations of technonature. Being cyborganic in these mediations of exchange, then, means much more than merely buying a suburban dwelling, an automobile, an airline ticket, or a television; but buying all of these does remediate cyborganized life forms toward more globalized forms of urbanatural being. Being cyborganic has different ontic qualities than living close to, down in, and covered with green-zoned ecologies still commonly called "Nature." These qualities are not, however, random chaotic motions of conflicting and colliding bodies; they are all tech-nonature's accidental normality at work.

Urbanatura, then, problematizes common conceptualizations about what the Earth's "environment," a national "economy," or any regional "city" is understood to be. As the "housing effects" exerted by "glassing over" or "gassing up" the Earth are rethought as the greenhouse effect of climate change, technonature should guide a foundational reimagination of the innovations unfolding at many of these intersecting trends in what are called "Nature" and "Society." Technonatural formations might be "complicitous and always opaque," but they also are "the best means for the global social order to extend its immanent and permanent rule to all individuals" (Baudrillard 1996, 196). Through "inventing the future," technonature's spatial practices set out boundary properties, whose prop-erty boundaries then frame their effects and ethics for the individuals and collectives they capture within the unending innovation that is the acci-dental normality of technonatural life.

Works Cited

The initial version of this chapter was presented at the annual meeting of the Association of American Geographers, March 7–11, 2006. Parts of this chapter previously appeared in *Organization and Environment* 18, no. 4 (2005) and *Environment and Planning A* 40 (August 2008).

Agger, B. 1989. *Fast Capitalism.* Urbana: University of Illinois Press.

Arendt, H. 1998. *The Human Condition.* Chicago: University of Chicago Press.

Armstrong, D., M. Burke, E. Lambert, N. Xardi, and R. Verry. 2002. "85 Innovations." *Forbes,* December 23, 124.

Bartsch, I., C. DiPalma, and L. Sells. 2001. "Witnessing the Postmodern Jeremiad: (Mis) Understanding Donna Haraway's Method of Inquiry." *Configurations* 9: 127–64.

Baudrillard, J. 1996. *The System of Objects.* London: Verso.

———. 1981. *For a Critique of the Political Economy of the Sign.* St. Louis, MO: Telos.

Castree, N. 1995. "The Nature of Produced Nature: Materiality and Knowledge Constitution in Marxism." *Antipode* 27, no. 1: 12–48.

Dodds, F., and T. Middleton. 2000. *Earth Summit 2002: A New Deal.* London: Earthscan.

Fischer, F. 1990. *Technocracy and the Politics of Expertise*. London: Sage.

Foucault, M. 1991. *The Foucault Effect: Studies in Governmentality*, ed. G. Burchell, C. Gordon, and P. Miller. Chicago: University of Chicago Press.

———. 1980. *History of Sexuality*, Vol. I. New York: Vintage.

Giedion, S. 1948. *Mechanization Takes Command: A Contribution to Anonymous History*. New York: Norton.

Habermas, J. 1991. *The Structural Transformation of the Public Sphere: An Inquiry into a Category of Bourgeois Society*. Cambridge, MA: MIT Press.

Haraway, D. 1991. *Simians, Cyborgs, and Women: The Reinvention of Nature*. New York: Routledge.

Hardt, M., and T. Negri. 2000. *Empire*. Cambridge, MA: Harvard University Press.

Hoffman, A. 2003. "Linking Social Systems to the Industrial Ecology Framework." *Organization and Environment* 16 (March): 66–86.

Jameson, F. 1991. *Postmodernism, or the Cultural Logic of Late Capitalism*. Durham, NC: Duke University Press.

Kelly, K. 1994. *Out of Control: The Rise of Neo-Biological Civilization*. Reading, MA: Addison-Wesley.

King, A. 1995. "Avoiding Ecological Surprise: Lessons from Long-Standing Communities." *Academy of Management* 20 (December): 961–65.

Lefebvre, H. 2003[1970]. *The Urban Revolution*. Minneapolis: University of Minnesota Press.

———. 1991. *The Production of Space*. Oxford: Blackwell.

———. 1984. *Everyday Life in the Modern World*. New Brunswick, NJ: Transaction.

Light, A., and B. Shippen, Jr. 2003. "Should Environmental Policy Be a Publicly Provided Good?" *Organization and Environment* 16 (June): 232–42.

Long, D. 2004. *Global Warming*. New York: Facts on File.

Luke, T. 2005. "The Death of Environmentalism or the Advent of Public Ecology." *Organization and Environment* 18 (December): 489–94.

———. 2001. "Real Interdependence: Discursivity and Concursivity in Global Politics." In *Language, Agency, and Politics in a Constructed World*, ed. Francois Debrix. Armonk: Sharpe.

———. 1999a. "From Body Politics to Body Shops: Individual and Collective Subjectivity in an Era of Global Capitalism." *Current Perspectives in Social Theory* 19: 91–116.

———. 1999b. *Capitalism, Democracy, and Ecology: Departing from Marx*. Urbana: University of Illinois Press.

Mumford, L. 1961. *The City in History: Its Origins, Its Transformations, and Its Prospects*. New York: Harcourt, Brace, and World.

Perrow, C. 1984. *Normal Accidents: Living with High-Risk Technologies*. New York: Basic.

Post, T. 2002. "85 Years and Ideas." *Forbes,* December 23, 123.

Shellenberger, M., and T. Nordhaus. 2004. "The Death of Environmentalism: Global Warming Politics in a Post-Environmental World." http://www .thebreakthrough.org/images/Death_of_Environmentalism.pdf.

Swyngedouw, E. 2004. *Social Power and the Urbanization of Water: Flows of Power.* Oxford: Oxford University Press.

Virilio, P. 2000. *A Landscape of Events.* Cambridge, MA: MIT Press.

———. 1997. *Open Sky.* London: Verso.

———. 1995. *The Art of the Motor.* Minneapolis: University of Minnesota Press.

Chapter Nine

Fluid Architectures: Ecologies of Hybrid Urbanism
Simon Guy

> Architecture and nature are not opposed, they are one and the same in differ-
> ent forms. The architect may select his path with more or less felicity, but his
> building blocks, in both literal and figurative sense, are the fruit of the earth.
> — Jodidio 2006, 6

> Our technological culture is one ecology, one of many possible ecologies.
> — Marras 1999, 5

Two ordinary stories: (1) Turning on my computer this morning to begin
writing this chapter, I raised the blinds and was greeted with sunshine
streaming through my windows — an event in Manchester in the north of
England where I live rare enough to be celebrated by flinging open the
front door to welcome in the warm air and light. As I began tapping away,
enjoying the feeling of connection to a benign nature, I was distracted by
a sudden blur of movement and looked up to find a grey squirrel that my
neighbour has been feeding and slowing taming staring back at me from
across my living room. We both paused for a moment before I instinc-
tively leapt out of my seat to chase the frightened creature out back
through my front door, back to where, I felt, it belonged. (2) While I was
sitting in a coffee bar on a hot day in Kyoto contemplating the lack of
pavement cafés in Japan and enjoying rare direct access to the river and a
cool breeze, my reverie was disturbed by the waitress, who was keen to
close the generous glass doors and cut off my connection to nature. The

ladies, she explained, were being disturbed by the insects. When I suggested that this would make us all very hot, she reassured me that she was also turning the air conditioning on.

Mundane, everyday experiences, but revealing of the ongoing socio-technical boundary work that shapes and literally "constructs" our experience of nature, buildings, and technology in a relational and contextual process. Maria Kaika has noted that this "double scripting of nature, [an almost] schizophrenic attitude towards both nature and the city, is central to modernist architectural visions and planning practices" (Kaika 2004, 14). Kaika charts the historical legacy of this separation of inside and outside, nature and culture, in the production of the bourgeois home. Tracing back historically to John Ruskin's notion of the home as a place of "shelter" and "peace" from all "terror, doubt and division," Kaika explores how the modern house becomes a modern home as an autonomous protected utopia "through a dual practice of exclusion: through ostracizing the undesired social as well as the undesired natural elements and processes" (ibid., 52). In this way, "unwelcome social and natural elements (from sewerage to homelessness) were exiled underneath or outside the modern home, below the streets and inside the walls, eliminated into underground passages, sent to a domain separate to that of dwelling places of the modern individual" (ibid.).

The logic of this technonatural boundary work is now being questioned in debates about sustainable architecture. Some wish to open the doors to "nature" to learn from, mimic, and experience nature in all its volatile forms. Others wish to strengthen the boundaries through smarter uses of technology so as to literally insulate us from the growing threats of increasingly unpredictable nature. Debates about sustainable building—in particular, the ways in which nature is translated and rescripted into design practices—graphically highlight how architecture has become a key "technonatural time/space," one that reveals this process of contestation over the meanings of nature (White and Wilbert 2006, 95). Buildings appear to powerfully symbolize our relationship to nature: according to some, they emphasize our deepening alienation from a previous state of harmony with nature; according to others, they provide shelter and comfort from a hostile nature. The result is some rather contradictory and confusing ideas about buildings and nature.

Of course, buildings *do* keep hostile elements at bay, and they *do* allow people to live and work in otherwise uninhabitable parts of the world.

More than that, they can be designed to provide a stable, comfortable, and utterly predictable indoor climate. However, contentment with technologies of shelter has become increasingly tempered by anxiety about their effects on our health and welfare and by knowledge of their contribution to global warming. Buildings and building technologies have become entwined with what George Myerson has termed the "ecopathology of everyday life [in which] there is no such thing as simply a blocked drain" (Myerson 2001, 52). Instead, "the blocked drain is a symptom of global environmental change, a mundane confirmation of a deeper meaning that has been discovered behind everyday life" (ibid.). No longer can we rely on purified images of inside and outside or simply advocate a return to nature. Institutional and cultural expectations have been transformed by, through, and with changing building technologies. Moving back in with the animals for the winter is simply not an option, for current lifestyles presume—and to a degree depend on—a standardized and well-managed indoor environment (Guy and Shove 1996, 2000).

Standing between the inside and the outside environment, buildings provide a tangible illustration of changing beliefs about the relationship between these two worlds. Exploring concerns about and responses to environmental change means acknowledging complex interrelations among the various environmental spaces we inhabit. Now viewed as a vulnerable domain, as well as a potentially inhospitable one, the outside world is managed more cautiously. Buildings are now being designed to use less energy and to make fewer demands on limited natural resources. One result is that the relationship between buildings and nature now evokes strong sentiments from environmentalists and architects, with buildings often becoming symbols of deference to *or* defiance of nature. Moreover, the diversity of responses to the choices architects make is quite bewildering, and rather than diminishing over time, that diversity seems to be increasing (Guy 2000; Guy and Farmer 2001; Guy and Moore 2005).

After more than four decades of debate about sustainable architecture, the search for some form of consensus around universal best environmental practices seems to have failed. The huge diversity of approaches to sustainable architecture defies any predefined technical or narrow ideological vision of sustainability. However, if viewed as technonatural constructs, these myriad design strategies serve as illuminating examples of the paradoxes that inevitably face efforts to design, develop, and promote sustainable buildings and cities. The rest of this chapter explores the plurality of design responses to the sustainability challenge, identifies the hegemonic framing and disciplining of architectural debates over sustainability, and offers an alternative account that argues for a more fluid interpretative frame that acknowledges and even celebrates design diversity.

Purity and Politics: sustainable architectures

There may well be as many types of relationships between nature and architecture as there are architects and buildings (Jodidio 2006, 7). A quick survey of any of the many illustrated reviews of sustainable architecture highlights the contested nature of environmental innovation. Such confusion is well illustrated in an edition of *Architectural Design* (Edwards 2001) that presented a set of "green questionnaires" completed by eminent architects—Norman Foster, Richard Rogers, Jan Kaplicky, Ken Yeang, and Thomas Herzog—who demonstrated these contradictory ways of seeing. Each architect was asked about his definition of sustainable design, his key concerns with regard to sustainability, how he would judge success in terms of sustainability, and how he used nature as a guide in his design work. For Kaplicky, the "major aspects of sustainable design are choice of materials and the performance of a building once it is built" (ibid., 34), whereas for Rogers sustainable design must include "a concern for the principles of social and economic sustainability as well as for the specific concerns of energy use and environmental impact of buildings and cities" (ibid., 36). For Herzog, it is about "using renewable forms of energy— especially solar energy—as extensively as possible" (ibid., 74), while for Yeang it is "design that integrates seamlessly with the ecological systems in the biosphere over the entire life cycle of the built system" (ibid., 60).

The relationship of architecture to nature is also contested. Rogers rather vaguely argues that "nature provides inspiration, information and analogy" (ibid., 36). Others are quite precise in linking natural and human processes. Yeang, for instance, believes that "nature should be imitated and our built systems should be mimetic ecosystems" (ibid., 60); similarly, Kaplicky feels that "there is much to learn from [nature's] more efficient use of materials" (ibid., 34). By contrast, Herzog does not believe that "architecture can be deduced directly from nature, since the design process and function of our buildings are quite different from what is found in most plants and animals" (ibid., 74), while Foster prefers to "look to human natures, vernacular traditions that are specific to the area in which we are working" (ibid., 32).

There are similar levels of disagreement over how we might recognize and assess architecture's success in becoming sustainable. Rogers contends that his practice meets the challenge of sustainability through "intelligent design and building fabric which contribute to a substantial reduction in running and maintenance costs during the life cycle of a building" (ibid., 36). Foster agrees, arguing that "a 'green' building will use as little energy as possible and will make the most of the embodied energy required to build it." He adds that it should also "create its own energy" and have structural flexibility in order for its life to be prolonged. Foster confidently suggests that his own Reichstag building has already "proved these concepts" (ibid., 32). Herzog concurs with this emphasis on "overall performance," but he also argues that only "beautifully made buildings" can really be sustainable and that architects must develop "new forms of architectural expression which are closely linked to the local micro-climate and topography, the natural resources and the cultural heritage of a certain region" (ibid., 74). Yeang goes further, insisting that a "successful 'green' building is only one that integrates seamlessly with the natural systems in the biosphere"; he then warns that "designers should also beware of making excessive claims about the sustainability of their designs because ecological design is still in its infancy" (ibid., 60). Kaplicky concurs with Yeang, arguing that as yet "there have been no truly green buildings built ... The buildings that are currently being constructed aren't even prototypes for a green age. They are only minor attempts at sustainability" (ibid., 34).

As even this small sample suggests, the mainstream of architecture is in some disagreement about design priorities, the role of technology, the importance of aesthetics, the relationship between natural and built environments, and the degree of optimism or pessimism that the current state of sustainable architectural practice should inspire. It is perhaps not surprising that given this complexity and potential for contradiction, Foster is tempted to define sustainable design as simply just "good architecture" (ibid., 32). However, it is again not surprising that Foster's rather optimistic view contrasts sharply with those of architects such as James Wines (2000), who want to emphasis that

virtually no form of shelter constructed today (with the exception of habitat built by a few remaining aboriginal cultures) can be credited as authentically green. Everything that technologically dependent societies assume is essential for survival—including the remedial solutions offered by the greenest of green architects—is plugged into the same diminishing sources of power. Every absorber plate and foil insulator required to build a solar collector, every chemical detergent used in a waste-composting plant, every ream of paper needed to spread the ecological message and every drop of jet fuel consumed in transporting environmentalists to international conferences places an additional drain on these resources. (2000, 226)

It appears, as Kenneth Frampton has put it, that there are as "many ways of practising [sustainable] architecture as there are architects" (2001: 128). That approaches to sustainable design are multiple and often contested is not an original claim. In fact, it has become rather commonplace to highlight two opposing and contradictory ways of valuing nature within the sustainable-architecture debate. Suzannah Hagan summarizes this perspective well: "At present, environmental architecture is split between an arcadian minority intent on returning building to a pre-industrial, ideally pre-urban state, and a rationalist majority interested in developing the techniques and technologies of contemporary environmental design, some of which are pre-industrial, most of which are not" (Hagan 2001, 4). Several other authors, including Farmer (1996), Pepper (1996), and Steele (2005), have developed ideological models or classification systems to account for the various approaches apparent within the sustainable-design debate. Though diverse, these models share a similar starting point in that they tend to recognize both the contested nature of the sustainability concept and the need to encompass the differing values of those individuals involved in the design process when understanding buildings.

Within the sustainable architecture literature this focus on ideology has tended to result in a relatively limited dualistic categorization of values in which the nature/culture relationship is portrayed as an expression of two very different but long-standing traditions. David Schlosberg summarizes: "The split between more traditional and conservative conservation groups and the more radical parts of the environmental movement today is simply a manifestation of basic differences between the utilitarian and romantic attitudes that began at the turn of the century" (Schlosberg 1999, 25). Similarly, David Pepper's well-known work on environmentalism identifies a dualistic debate that questions whether "green strategies" should follow an ecocentric ("radical") approach or a technocentric ("reformist") approach to tackling environmental problems (Pepper 1996, 7). Put simply, technocentrics adhere to a process of "ecological modernization that] indicates the possibility of overcoming the environmental dilemma without leaving the path of modernization" (Spaargaren and Mol 1992, 334), whereas ecocentrics believe that a radical new way of living is the only way forward if we are to avoid the looming ecological crisis.

This technocentric versus ecocentric debate is reflected in debates around sustainable architecture where for many, "technology remains the answer to saving the environment," while others "argue that technology ... is the primary cause of destruction of nature, and that expecting it to provide a solution for environmental ills is like using the cause of the disease to cure it" (Steele 2005, 291). Seen this way, the search for sustainable architecture is a struggle over facts and values in an ongoing process of renegotiations among design, technology, and nature.

Valuing Architecture/Nature

Environmental architecture, in other words, is environmental architectures— that is, a plurality of approaches, with some emphasizing performance over appearance, and others appearance over performance. (Hagan 2001, 4)

Most policy responses to the fluid interpretations and practices of sustainable architecture express a concern over the lack of precision and quantification. It is, again, commonplace to suggest that if we are to achieve sustainable buildings, then architecture should become more "objective [and that] until a consensus is attained, the ability of the architectural community to adopt a coherent environmental strategy, across all building types and styles of development, will remain elusive" (Brennan 1997, 25). So while Pepper characterizes environmental value historically through a prism of shifting attitudes toward technology, more recent policy and practice perspectives have developed categories based on the notion of differential environmental performance. As Hagan has argued, "it is the rationalist majority who now dominate the field. One has only to look at the proceedings of any conference on environmental architecture in the last twenty years to see the overwhelming emphasis on the scientific and quantitative dimensions of the discipline: thermal conductivity of materials, photovoltaic technology, computer simulations, life cycle analysis, and so on" (Hagan 2001, x).

And the process of reclassification doesn't stop here. Some suggest that a distinction be made between "green" and "sustainable," with green buildings defined as those that achieve incremental improvements in performance relative to typical practice. Sustainable buildings, on the other hand, would be those that achieve a more radical "absolute" performance measured against global "biosphere health" and "carrying capacity criteria" (Cole 1999, 232–33). A further distinction is provided by Cole, who calls for different shades of green: "A deep green building may, for example, refer to one designed from the outset to maximize the use of solar energy, daylighting and natural ventilation, as well as harvest rainwater, treat any wastes on site and use environmentally sound materials in the most efficient way. Light green, by contrast, may refer to buildings that have incorporated one or more green features such as high-efficiency windows, high recycled-content carpets or automatic shut-off systems for lights but are otherwise conventional" (ibid., 233). Haughton's similar categorization would depend on a building's ability to close the circuits of resource supply and waste. A deep-green building would have a circular metabolism, while a light-green building would have a linear albeit reduced metabolism (Haughton 1997, 189–95).

This interpretive frame, which Macnaghton and Urry term "environmental realism," is founded on the notion that "rational science can and will provide the understanding of the environment and the assessment of those measures which are necessary to rectify environmental bads" (Macnaghton and Urry, 1998, 1). The result is a dominating focus on reducing the energy intensity of buildings — for example, through the use of insulating materials, low-energy lighting, and natural ventilation and by eschewing non-renewable and potentially toxic materials. Energy economics is a major priority among these practitioners. As Susan Maxman argues, "it's not like the 1970s, when every house had to be earth-bermed, solar powered, etc. ... We realize now that it has to make economic sense as well" (Maxman 1993, 11).

This popular view of sustainable architecture renders it roughly synonymous with energy efficiency. In Britain, for example, Brian Edwards and Paul Hyatt have written a "rough guide to sustainability," published by the Royal Institution of British Architects (RIBA), in which they confidently link sustainability to "a number of important world congresses" through which we have learned what it means to be sustainable (Edwards and Hyatt 2001, 1). Edwards and Hyatt link architectural sustainability to the much quoted Brundtland definition by emphasizing the limits to the "carrying capacity" of the planet, and they point to Britain's Building Services Research and Information Association (BSRIA) definition of sustainable construction: "The creation and management of healthy buildings based upon resource efficient and ecological principle."[1] Drawing from these sources, they argue that a "large part of designing sustainably is to do with energy conservation," but they also recognize that it is also about "creating spaces that are healthy, economically viable and sensitive to local needs" (ibid., 7). However, the rest of their guide focuses almost exclusively on resource efficiency and has little to say about wider social and political issues. They claim that these alternative visions of how we might best live in harmony with nature can be adequately expressed through an energy-rating model.

From an American perspective, Harry Gordon maintains that the "LEED standards, issued in 2000, are creating a common understanding of what it means to build green" (Gordon 2000, 34). Employing similar logic, Paul Hawken, Amory Lovins, and Hunter Lovins, in their highly influential book, *Natural Capitalism*, argue that consumers will automatically embrace radical resource efficiency once they understand that they can reduce consumption "without diminishing the quantity or quality of services that people want" (Hawken, Lovins, and Lovins 1999, 176). This level of self-confidence in the compelling transparency of sustainable architecture to produce social and environmental change assumes a purely scientific or quantitative framing of the problem and that there are no barriers—save our awareness—to implementation (Guy and Shove 2000).

But as Mark Wigley has suggested, the "overt politics of ecology—the equitable management of resources—almost always preserves certain regressive ideological formations" (Wigley 1999, 48). The "technocist supremacy" that dominates most environmental-research programs often results in "non-technical" issues—of health, poverty, and equity, for example—being downplayed or ignored (Woodgate and Redclift 1998). Implicit in this model of consensus is a "process of standardization" whereby "particular local conditions" and competing "forms of local knowledge" tend to be ignored (ibid., 9). The production of best technical practices emerges from an interpretive squeeze in which problems amenable to technical fixes are foregrounded at the expense of a range of other environmental concerns (Guy and Farmer 2001). The recent push to define and codify of zero-carbon buildings in the United Kingdom is the next step toward a "single national standard" recognizable through a quantitative assessment of environmental performance in buildings.[2] Of course, not every architect is satisfied with this project of ecological modernization.

Environmental Icons

James Wines has led the response to the "technologizing" of debate, policy, and practice around sustainable architecture. As he puts it,

> Certainly one of the major problems facing environmental architecture, aside from the absence of a strong societal endorsement, is a professional choice to over-emphasize the technological advantages and undervalue the social and aesthetic aspects. There is an unbalanced amount of effort currently being spent to create a sanctimonious mythology around what is basically a collection of admirable engineering innovations. These techno-remedial solutions — often promoted by architects as challenges to public conscience — tend to discourage potential sympathizers by their reproachful tone and the sententious way the message is delivered. Although people are rather fascinated by end-of-the-world scenarios and fantasy cures, there is nothing particularly compelling about technical reports on photovoltaic cells, solar panels, and thermal glass, all of their admirable green intentions notwithstanding. (2000, 64)

For Wines, the role of architectural design is to develop an "environmentally based imagery, an eco-centric technology, and a corresponding philosophy" (Wines 1997, 113). He contends that the stakes are very high. The result of focusing so narrowly on technology, he feels, has been to allow "civilisation's connection to nature [to] die in the mind" (ibid.). A "new sense of fusion with nature and attention to the interactive elements joining disparate parts challenge every green architect today" (ibid., 114). Charles Jenks has championed this view in his celebration of postmodern "fractal" architecture, declaring that it is "essential to cultivate a tradition of sensuous, creative Green architecture — one based on the new science of complexity" (Jenks 1997, 96). He is content to leave "political and energy issues to others" in order to "concentrate on an architecture that delights in the ecological paradigm for its philosophy of holism, its style and the way it illuminates the complexity paradigm" (ibid., 94). From the perspective of both Wines and Jenks, the role of sustainable architecture is to break free from architectural ideologies that reflect a humanist culture — one that takes an anthropocentric attitude toward nature. According to Wines the "entire direction in design suggests the development of a new paradigm in the building arts that are based on ecological models" (Wines 1997, 33).

In *Designs for a Real World*, the eco-architect Victor Papenek presents a range of innovations drawn from what he terms the "handbook of nature," arguing that through "analogues to nature, man's problems can be solved optimally" (Papenek 1971, 186). But as Andrew Ross has argued, "technologies based on natural elements, or imitating natural processes are no guarantee of health, sustainability, or even biodiversity" (Ross 1994). Instead we are presented with a symbolic vision of environmentalism in which, as Mark Wigley suggests, "ecology is a question of images in the end, images of architecture and the architecture of images" (Wigley 1999, 27). Wigley further argues that "such images are rarely innocent" (ibid.). As with the technologizing of sustainable architecture discussed above, this aesthetic practice of sustainable architecture works by privileging particular (in this case aesthetic/symbolic) environmental concerns while downplaying and even erasing a wide range of broader issues and, in particular, political questions.

Jenks's book, for instance, is filled with sensuous images of his private estate in Scotland, with the carefully cultivated lawns resembling an updated scientific vision of the landed estates of the eighteenth century. While Jenks may rightly argue that "good ecological building may mean bad expressive architecture" (Jenks 1997, 94), it is equally true, as Ross again points out, that "biomorphic houses designed according to botanical principles... are just as fitting instruments of social apartheid as housing modelled on industrial factories (Ross 1994, 271).

Looking back, in terms of the commitment to sustainable architecture, "best practice" technical performance and iconic symbolism seem to represent flip sides of the same coin. Each seeks to renegotiate our relationship with nature, and both are equally committed to rewriting and then confirming universal stories about architecture and sustainability. Both are involved in strongly editing and translating debates about sustainability and then inscribing the resulting practices through the (re)design of buildings, landscapes, and cities. Contestation over the meaning of sustainable architecture is then less the result of uncertainty and more a consequence of the existence of what John Hannigan terms "*contradictory certainties:* Severely divergent and mutually irreconcilable sets of convictions both about the environmental problems we face and the solutions that are available to us" (Hannigan 1995, 30).

Critically, we should not conclude from this interpretive flexibility either that the sustainability debate is so mired in confusion that it should be abandoned, or that it simply represents a narrow ideology that mystifies the challenges facing contemporary architecture. This latter view has been mobilized by Audacity, a firm that describes itself as a "campaigning company concerned with the design and production of the man-made environment, advocating development free from the burden of "sustainababble" and "communitwaddle.[3] For Audacity founder Ian Abley, the real issue is one of "failing to raise the level of machine age industrial development around the world" and instead pandering to "advocates [of] ever more restraint and regulation" (Abley and Heartfield 2001, 19). Abley argues for a renewal of industrial modernism to "produce our way out of fossil-fuelled site based construction" in order to avoid what he terms an "increasingly thin architectural façade over a stifling parochialism morally justified as sustainable development" (ibid., 16).

The tone of Abley's co-edited collection (Abley and Heartfield 2001), notably titled *Sustaining Architecture in the Anti-Machine Age*, is exemplified in architecture journalist Austin Williams's chapter, which similarly argues the need to "challenge the current positive perceptions about the religion of 'sustainability' in architecture" (Williams 2000, 49). This position is expressed in its most extreme form by fellow Audacity contributor Martin Pawley, who connects current environmental concerns with those of the Nazis, who had quite a track record for environmental innovation: "Surely only lack of historical knowledge saves these advocates of sustainability from being recognised for the Nazi fellow travellers they are" (Pawley 2003). The fear here is of a modernist project thrown in doubt and of a threat to free-market capitalism, which has sponsored and profited from (more or less) unregulated development for so long. Of course, Charlick and Nicholson, in the same collection, are right to warn that "much sustainable architecture deals in symbolism" and that the "green roof or the solar array represents a 'green' sensibility, which may extend no further" (Charlick and Nicholson 2001, 68). Moreover, while they also signal the loss of an "unviolated" or "idyllic" "first nature," they see hope in the production of a "second" or "hybrid nature" in which the boundaries between nature and technology are blurred (ibid., 66).

Building Hybridity

> What is clear is that there is no still point of the turning world as far as green
> is concerned. Variations are thrown up by social, political, cultural and eco-
> nomic factors, as well as by individual preferences. (Castle 2001, 5)

In sum, exploring debates and mapping practices of sustainable architecture involves tracing the interplay of competing environmental values
and practices through the enactment of alternative design logics as they
shape the technonatural profiles of green building development (Guy and
Farmer 2001; Guy and Moore 2005). For as Noel Castree has put it, and as
practices of sustainable architecture make clear, "ideas about nature" do
not "somehow touch down uniformly across time and space. Rather they
are produced by myriad knowledge-communities who possess similar
(and sometimes different) outlooks on nature" (Castree 2005, xiv).

Seen this way, alternative design strategies are the result not simply of
contestation over technological optimization or expression of the environmental sublime, but also of distinct philosophies and practices rooted
in differing accounts of the nature/culture relationship. In order to more
fully understand the heterogeneity of sustainable architecture, we therefore have to account for the multiple ways in which environmental problems are identified, defined, translated, valued, and then embodied in
built forms through diverse design and development pathways. The current "society/nature dualism" that, as we have seen, is structuring debate
about sustainable architecture is, as Castree again suggests, blinding us to
the "need for a new vocabulary to describe the world we inhabit" (ibid.,
224). For Castree this would not be a "vocabulary of 'pure forms' — in
architectural terms, the 'performative'" or the "iconic" — but one that
"captures the hybrid, chimeric, mixed-up world in which we are embedded" (ibid., 224).

From this perspective it is clear that we need to open up and explore
the language we use to talk about sustainable architecture. As Andrew
Jamison has suggested, "more fluid terms are needed: dialectical, open-
ended terms to characterize the ebbs and flows, nuances and subtleties
and the ambiguities of environmental politics" (Jamison 2001, 178).
Similarly, David Schlosberg has called for "statements that are open
rather than doctrinaire," that "conscript" rather than alienate, and that
encourage a debate in which "discourse is never-ending, and solidarity

is forever creating new networks and mosaics" (Schlosberg 1999, 103).

Seen this way, we need more fluid interpretations of sustainable architecture. This is not to suggest that buildings are infinitely flexible, subject only to the whims of designers. The obduracy of certain materials and the contingencies of particular technologies are part of the story of building design and development, as Annique Hommels has shown (Hommels 2005). Rather, fluidity here suggests an interpretive flexibility and plasticity of design and technology, or as Steven Moore has put it, "it could have been designed a different way" (Moore 2001, 25). Following the fluidity of design, we would ask: How and why are designers pursuing environmentalism in very particular ways, with very different notions of nature and culture, and with highly variable technological strategies?

Critically, our use of the term "technology" here is an expansive one. We mean by it not only the artifacts associated with sustainable architecture — solar collectors, wind generators, biomass boilers, and the like — but also the knowledge required to construct and use these artifacts, as well as the practices that engage them. This stance echoes that of Andrew Feenberg, who has similarly explored these approaches and emphasized the need to avoid the essentialist fallacy of splitting technology and meaning and to focus instead on the "struggle between different types of actors differently engaged with technology and meaning" (Feenberg 1999, xiii). Seen this way, the contexts of technology include such diverse factors as "relation to vocations, to responsibility, initiative, and authority, to ethics and aesthetics, in sum, to the realm of meaning" (xiii). Wrapped up in each technological artifact — or, in the case of our architectural interests, each building — is an assemblage of ideologies, calculations, dreams, political compromises, and so on. Similarly, Eric Swyngedouw has pointed to the "combined metabolic transformations of socio-natures" in the construction of a skyscraper, which testifies "to the particular associational power relations through which socio-natural metabolisms are organized (in terms of property ownership regimes, production or assemblaging activities, distributional arrangements and consumption patterns)" (Swyngedouw 2006, 109).

Tracing these networks would mean looking beyond the polemical debates about architectural visions and sustainable futures to the often messy ways in which architectural artifacts are assembled on local sites, are funded by particular financial regimes, utilize specific expertise (or lack thereof), connect to technical networks, are argued over by a restricted or enlarged community of users, and are placed within a planning framework (or not). Again, looking beyond the ideological choice between employing the technical disciplines of energy management to produce new forms of "smart buildings," or simply mimicking of organic nature in architectural form, dialogue about sustainability might come to inhabit what Amerigo Marras has termed an "architecture of the inbetween" (Marras 1999). For Marras, design involves a weaving of ecology and technology in a "transformative flux...a catalytic fusion...intentionally generating some hybrid transgendering paradigms" (ibid., 3). Rejecting the "extreme positions of being either-or," Marras urges the "fluid process of in-between" (ibid., 6). Adopting this way of seeing and describing building design as "fluid," we might better recognize both the hybrid nature of the green building and competing pathways toward sustainable futures.

Tracing Fluid Architectures

Tracing these fluid hybrids means looking beyond fixed definitions and dualistic typologies, while at the same time resisting the temptation to either abandon the environmental project (a.k.a. Audacity) or simply swim along in a ocean of free-flowing design options with no fixed reference points. It also means neither accepting the status quo—familiar buildings symbolically retrofitted with wind turbines and solar collectors—nor exclusively searching for radically new typologies. Looking back across the competing definitions of sustainability offered by the leading architects reviewed above, we might be unable to identify any semantic solutions to what sustainability really means in architectural terms, but we can find a convincing and workable tool box of design innovations, technological options, and creative practices. The question is less whether any combination of these might provide a universal blueprint (they won't), and more how they might contribute to meeting specific environmental challenges. Seen this way, we might begin to sketch out some general principles of "fluidity" in order to frame diverse sustainable-design approaches that may aspire to be flexible, situated, pragmatic, *and* participative. We can very briefly (and tentatively) explore what we might call these design frames and illustrate them with examples of architecture beyond the fold of what is conventionally thought of as sustainable architecture.

First, "flexibility" to a range of technological options—be they high-tech or low-tech—and an appetite to mix these where it makes sense (more of which below). Equally, looking beyond contested league tables of environmental performance in terms of materials (wood versus concrete), height (skyscrapers versus groundscrapers versus underground architecture), location (cities versus suburbs versus rural villages), and a willingness to be open to heterogeneous combinations of purpose and program, from "mixed" to "mixed up" uses. Echoing the emphasis on "interpretive flexibility" found in Science and Technology Studies (cf. Bijker 1995), the point here is not to abandon judgment but to avoid closing down the evaluative process prematurely, to always be open to other design possibilities.

One exemplar here is the work of Atelier Bow-Wow in Japan, which focuses on the narrow, in-between spaces of Tokyo, where uses are constantly shifting. Bow-Wow has made a study of what its members term Tokyo's "pet architecture" — that is, buildings that appear monstrous in their rejection of standardized design and purpose and that instead celebrate wild juxtapositions of use: temples and shops, laundries and saunas, shrines and restaurants, pachinko parlours and banks, taxi companies and golf driving ranges.[4] Bow-Wow uses these pet architectures as an inspiration for its own design practices, which are characterized by strategic interventions into the existing fabric in response to new demands, through creative conversion and adaptation of the built fabric of Tokyo. While not overtly a "green" practice, Bow-Wow has dedicated itself to satisfying changing human needs, intensifying use of urban space with great economy and efficiency, and focusing on recycling and reusing space.

So far, so fluid. What is needed, second, is a frame that will give shape to this fluidity — a frame we might find in design that is "situated." This familiar architectural trope is often promoted in terms of "regionalism" (cf. Kenneth Frampton's critical regionalism) and is familiar to environmentalists in discourses of "bioregionalism." But here the emphasis is away from fixed spatial containers — local, city, region, defined simply and essentialized by their cultural and/or physical characteristics — and more on creating situationally specific solutions to locally defined challenges. These local definitions might vary hugely, from comfort to community and from energy security to emergency shelter to flood prevention. To take this latter example, with climatologists predicting that precipitation in the Netherlands could increase by as much as 25 percent over the next few years, Dutch architects are now designing living accommodation, greenhouses, parking lots, and factories that can float and that could grow in the future into "waterproof" towns. In a recent example, the Dutch architecture practice Dura Vermeer has built twenty-six amphibious homes in Massbommel in the Netherlands, each built on a hollow concrete-cube base, which in turn is anchored to the land by a single vertical pile.[5] All utilities, including electricity and water, are brought into the house through flexible pipes that allow each house to adapt to a thirteen-foot rise in the water table. This is a response to a perennial challenge for Dutch urbanism; even so, it is not difficult to think of further adaptations of this "situated" response to environmental change, from New Orleans to Bangladesh.

Third, while environmental change might effect and be effected by us all, our strategies for ameliorating and coping with its causes and symptoms could vary dramatically. Which brings us to "pragmatism." Here we might follow Richard Rorty (who in turn is following John Dewey) when he calls on us to abandon "the attempt to find a (*single*) theoretical frame of reference within which to evaluate proposals for the human future" (Rorty 1998, 20). Instead, the pragmatic imperative is to deal with the particular challenge at hand. This is well illustrated in the work of Architects for Humanity, a charity-based practice headquartered in California but operating as a worldwide volunteer network. A declared principle for this work is "pragmatism" — specifically, providing shelter after disaster and for communities in need.[6] A typical project was the provision of global-village shelters in Grenada after Hurricane Ivan in 2004. The structures, which were made from recycled corrugated cardboard impregnated with fire retardant and laminated for water resistance, provided speedy transitional shelter that could be distributed and erected easily and quickly. Later refinements to this design led to the erecting of almost five hundred shelters in Pakistan following the Kashmir earthquake of 2005.

This perspective would emphasize, fourth and finally, a "participative" approach to design in which voices beyond the architect/developer/investor nexus are heard and make a difference. This frame encompasses notions of participatory politics (Barber 1984), but it also takes inspiration from Bruno Latour's concern in his *Politics of Nature* to multiply the number of representations of any specific issue (Latour 2004a). More specifically in design terms, it means building with the participation of the community you are building *for*. Sam Mockbee's Rural Studio practice exemplifies this approach. Its members talk about "sharing the sweat" with the community; they prefer to see themselves as "citizens" rather than experts.[7] Refusing to abstract "architect" from their context of "work," they aim for low-cost, sustainable, "workable solutions" for the economically disadvantaged local community in Alabama. Their 2001–2 "Lucy's House," designed for a six-person family previously occupying two rooms, was sponsored by Interface, the world's largest manufacturer of carpet tiles, which provided 72,000 individual stacked non-recyclable scrap tiles, which were then held in compression by a heavy wooden ring-beam from the carpet manufacturing process. In the resulting house, a tower form contains the parents' bedroom, with a spectacular view, atop a cast-concrete family room that performs a dual role as a tornado shelter. Rural Studio's approach merges the flexible, situated, pragmatic, and participative principles of fluid architecture with an adaptive assembly fit for a purpose and a community, both low-cost and low-impact.

Thinking through sustainable architecture from this "fluid" perspective does not mean rejecting one particular typology (skyscrapers) and celebrating another (vernacular). But it may mean valuing different aspects of the design. While the examples above were deliberately chosen to stretch our notions about what green architecture might represent, we can equally well reread some of our current icons of ecodesign within this different interpretive frame. Take Frankfurt's celebrated Commerzbank, a green skyscraper by Norman Foster. Typically the Commerzbank tower is celebrated as an ingenious feat of heroic design employing innovative technologies (which it is). But there is another, less known story about the Commerzbank: its success is seen not simply in terms of its environmental performance, but also in relation to how the design achieved a material resolution of competing political views of the future of the city. By designing a new ecological skyscraper, Foster and his supporters were able, through a process of public dialogue, to mediate a dispute between images of American (high-rise) and European (low-rise) style urbanism that encouraged both business groups and environmental campaigners to collectively embrace a more sustainable development pathway for Frankfurt (Guy and Moore 2007).

In sum, the effect of this frame is to change our understanding of the meaning of progress. Returning to Rorty, we might argue that "instead of seeing progress as a matter of getting closer to something specifiable in advance, we see it as a matter of solving more [local] problems" one at a time (Rorty 1998, 28). In design terms this translates as reducing our dependency on prepackaged, universalized design solutions and beginning each project with a process of identifying and prioritizing the key challenges to be tackled (Guy and Moore 2007).

In Conclusion — the Fluidity of Sustainable Architecture

> The old dreams of Eden still exist, and today's architects can conceive of "living" buildings, or a new, artificial nature — a world in which architecture and nature will become one. (Jodidio 2006, 28)

By exploring the social and technical fluidity of sustainable architecture, this chapter has sought to link itself to White and Wilbert's identification of a "new mood or sensibility, a shift in the imaginative horizons of the environmental debate" (White and Wilbert 2006, 100). In doing so, it has aimed to counter a prevailing technological pessimism about sustainable architecture (often echoed in "ecocentric" discourses) without uncritically celebrating the libratory potential of technology (often echoed in "ecotechnic" discourses), by re-emphasizing the ambivalence that is inherent in all technological innovation. The argument, echoing that of Bruno Latour, is that we should focus on matters of "concern" instead of fetishizing matters of "fact" and that in doing so we should seek "controversy" instead of searching for consensus (Latour 2004b). Instead of working to close down debate and squeeze out alternative design approaches, we should highlight the contradictory (dialectical, ambivalent, open, partially undetermined ...) nature of sustainable architectures. In this way we might counter the prevalent view of technological systems, such as architecture, which is, that they are not manageable tools but rather beyond our control and overwhelming us, resulting in resignation and pessimism that is ultimately disabling both intellectually and politically (Coutard and Guy 2007). By actively searching for and exploring contingency and contestation over urban technologies, we might begin to identify a politics of architecture and urbanism that breaks free from technological pessimism and that offers some hope for change.

This approach echoes that of David Schlosberg, who argues that "pluralism demands engagement" (Schlosberg 2006, 149). Critically, for Schlosberg, the "dilemma" of difference will not be overcome simply by liberal tolerance, but by what he calls "agonistic respect," which makes it tactically possible for those who may hold thoroughly allergic metaphysical beliefs—deep ecologists and ecological modernists, for example—to act together in achieving a particular limited goal, if not a totalizing utopian order (Schlosberg 1999, 4). By responding to Schlosberg's call to "acknowledge" and "recognize" the diversity of practices that might point to alternative sustainable futures, we may begin to chart an agenda for future research that will challenge the current orthodoxies and "engage" in the making and remaking of sustainable architectures (Guy and Moore 2005, 2007). In the process, debate about and practices of sustainable architecture would enter into what Hinchcliffe and Whatmore describe as a "reconfiguration of ecology, away from statements of fact to engagements with possibilities" (Hinchcliffe and Whatmore 2006, 131).

"Instead of unveiling the truth of the ecology at hand, there is a turn to those involved in the co-fabrication of living cities. There is a redistribution of expertise, or a redefinition of expertise so that it includes lay engagements with place" (ibid.).

The result, they argue, would be a turn away from the "clearly present" toward "experimentation in urban ecologies, and from a closed politics of ecological states of nature, towards a more open politics of things, of living as others." Such a position could be reflected in debates about sustainable architecture, for as Philip Jodidio argues: "Uncertainty, ambiguity, and a constantly evolving vision of just what nature is will guide architecture as long as there are buildings" (Jodidio 2006, 28).

Notes

1 See: http://www.bsria.co.uk.
2 See file://Users/simonguy/Desktop/Planning%20Portal%20-%20Code%20for%20Sustainable%20Homes.webarchive.
3 http://www.audacity.org/index.htm.
4 See Tokyo Institute of Technology Tsukamoto Lab and Atelier Bow-Wow, Pet Architecture Guide Book (Tokyo: World Photo Press, 2001).
5 See http://www.spiegel.de/international/spiegel/0,1518,377050,00.html (accessed 1/8/08).
6 See http://www.architectureforhumanity.org. Also see Architecture for Humanity (eds.), *Design Like You Give a Damn: Architectural Responses to Humanitarian Crises* (London: Thames and Hudson).
7 See http://cadc.auburn.edu/soa/rural%2Dstudio.

Works Cited

Abley, I., and J. Heartfield. 2001. *Sustaining Architecture in the Anti-Machine Age*. London: Wiley-Academy.

Barber, B. 1984. *Strong Democracy: Participatory Politics for a New Age*. Berkeley: University of California Press.

Bijker, W.E. 1995. *Of Bicycles, Bakelites, and Bulbs: Toward a Theory of Sociotechnical Change*. Cambridge, MA: MIT Press.

Brennan, J. 1997. "Green Architecture: Style over Content." *Architectural Design* 67, nos. 1–2: 23–25.

Castle, H. 2001. "Editorial — Green Architecture." *Architectural Design* 71, no. 4: 5.

Castree, N. 2005. *Nature*. London: Routledge.

Charlick, P., and N. Nicholson. 2001. "Ecological Frequencies and Hybrid Natures." In *Sustaining Architecture in the Anti-Machine Age*, ed. I. Abley and J. Heartfield. London: Wiley-Academy. 64–71.

Cole, R. 1999. "Building Environmental Assessment Methods: Clarifying Intentions." *Building Research and Information* 27, nos. 4–5: 230–46.

Coutard, O., and S. Guy. 2007. "STS and the City: Politics and Practices of Hope." *Science, Technology, and Human Values* (forthcoming).

Edwards, B., guest ed. 2001. "Green Architecture." In *Architectural Design* 71, no. 4: 5. London: Wiley-Academy.

Edwards, B., and P. Hyatt. 2001. *Rough Guide to Sustainability*. London: RIBA.

Farmer, J. 1996. *Green Shift: Towards a Green Sensibility in Architecture*. Oxford: WWF.

Feenberg, A. 1999. *Questioning Technology*. London: Routledge.

Frampton, K. 2001. "Technoscience and Environmental Culture: A Provisional Critique." *Journal of Architectural Education* 54, no. 3: 123–29.

Gordon, H. 2000. "Sustainable Design Goes Mainstream." In *Sustainable Architecture: White Papers*, ed. D.E. Brown, M. Fox, and M.R. Pelletier. New York: Earth Pledge Foundation.

Guy, S. 2002. "Sustainable Buildings: Meanings, Processes, Users." *Built Environment* 28, no. 1: 5–10.

Guy, S., and G. Farmer. 2001. "Reinterpreting Sustainable Architecture: The Place of Technology." *Journal of Architectural Education* 54, no. 3: 140–48.

Guy, S., and S. Moore. 2007. "Sustainable Architecture and the Pluralist Imagination." *Journal of Architectural Education* 60, no. 4: 15–23.

———, eds. 2005. *Sustainable Architectures: Cultures and Natures in Europe and North America*, Oxford: Spon.

Guy, S., and E. Shove. 2000. *A Sociology of Energy, Buildings, and the Environment*. London: Routledge.

———. "From Shelter to Machine: Remodelling Buildings for a Changing Environment." Proceedings of the World Conference of Sociology, Biederfeld, Germany, July.

Hagan, S. 2001. *Taking Shape: A New Contract Between Architecture and Nature.* Oxford: Architectural Press.

Hannigan, J. 1995. *Environmental Sociology: A Social Constructivist Perspective.* London: Routledge.

Haughton, G. 1997. "Developing Sustainable Urban Development Models." *Cities* 14, no. 4: 189–95.

Hawken, P., A. Lovins, and H. Lovins. 1999. *Natural Capitalism.* Boston: Little, Brown.

Hinchcliffe, S., and S. Whatmore. 2006. "Living Cities: Towards a Politics of Conviviality." *Science as Culture* 14, no. 2: 123–38.

Hommels, A. 2005. *Unbuilding Cities: Obduracy in Urban Sociotechnical Change.* Cambridge, MA: MIT Press.

Jamison, A. 2001. *The Making of Green Knowledge: Environmental Politics and Cultural Transformation.* Cambridge: Cambridge University Press.

Jenks, C. 1997. *Architecture of the Jumping Universe.* London: Wiley-Academy.

Jodidio, P. 2006. *Architecture: Nature.* London: Prestel.

Kaika, M. 2004. *City of Flows: Modernity, Nature, and the City.* New York: Routledge.

Latour, B. 2004a. *Politics of Nature: How to Bring the Sciences into Democracy.* Cambridge, MA: Harvard University Press.

———. 2004b. "Why Has Critique Run Out of Steam? From Matters of Fact to Matters of Concern." *Critical Inquiry* 30, no. 2: 225–48.

Macnaghton, P., and J. Urry. 1998. *Contested Natures.* London: Sage.

Marras, A., ed. 1999. *ECO-TEC: Architecture of the Inbetween.* New York: Princeton Architectural Press.

Maxman, S. 1993. "Shaking the Rafters." *Earthwatch*, July–August, 11.

Moore, S.A. 2001. *Technology and Place: Sustainable Architecture and the Blueprint Farm.* Austin: University of Texas Press.

Myerson, G. 2001. *Ecology and the End of Postmodernism.* Cambridge: Icon.

Papanek, V. 1971. *Design for the Real World: Human Ecology and Social Change.* New York: Pantheon.

Pawley, M. 2003. "Tracing the Roots of Sustainability to a Less-Than-Glorious Past." *Architects Journal*, April 3. http://www.audacity.org/MP-04-03-2003.htm.

Pepper, D. 1996. *Modern Environmentalism: An Introduction.* London: Routledge.

Rorty, R. 1998. *Achieving Our Country: Leftist Thought in Twentieth-Century America.* Cambridge, MA: Harvard University Press.

Ross, A. 1994. *The Chicago Gangster Theory of Life: Nature's Debt to Society.* New York: Verso.

Schlosberg, D. 2006. "The Pluralist Imagination." In *The Oxford Handbook of Political Theory*, ed. J. Dryek, A. Philips, and B. Honig. Oxford: Oxford University Press.

————. 1999. *Environmental Justice and the New Pluralism: The Challenge of Difference for Environmentalism*. Oxford: Oxford University Press.

Spaargaren, G., and A. Mol. 1992. "Sociology, Environment, and Modernity: Ecological Modernisation as a Theory of Social Change." *Society and Natural Resources* 5: 323–44.

Steele, J. 2005. *Ecological Architecture: A Critical History*. London: Thames and Hudson.

Swyngedouw, E. 2006. "Circulations and Metabolisms: (Hybrid) Natures AND (Cyborg) Cities." *Science as Culture* 15, no. 2: 105–22.

Williams, A. 2001. "Zen and the Art of Life-Cycle Maintenance." In *Sustaining Architecture in the Anti-Machine Age*, ed. I. Abley and J. Heartfield. London: Wiley-Academy. 42–51.

White, D., and C. Wilbert. 2006. "Introduction: Technonatural Time-Spaces." *Science as Culture* 14, no. 2: 95–104.

Wigley, M. 1999. "Recycling, Recycling." In *ECO-TEC: Architecture of the Inbetween*, ed. A. Marras. New York: Princeton Architectural Press. 38–49.

Wines, J. 2000. *Green Architecture: The Art of Architecture in the Age of Ecology*. Cologne: Taschen.

————. 1997. "Passages: The Fusion of Art and Architecture in the Recent Work of SITE." *Architectural Design* 67: 33.

Woodgate, G., and M. Redclift. 1998. "From a Sociology of Nature to Environmental Sociology." *Environmental Values* 7: 2–24.

Chapter Ten

A Post-Industrial Green Economy: The New Productive Forces and the Crisis of the Academic Left

Brian Milani

Technonatures takes off from an apparent malaise of the environmental movement. This chapter argues that a bigger problem may be a failure of the political and intellectual left to appreciate the relationship between human-development potentials and strategies for social change. I argue in this chapter that because we are in the midst of a major historical transition, it is important for us to go a step beyond conventional understandings of the relationship between humanity's "productive forces" and "relations of production."

This chapter takes a classical Marxist insight — that outdated social power relationships act as brakes on emerging productive forces — and applies it to a post-industrial context. It looks at the essence of post-industrialism — that is, its potential for facilitating a different kind of human and green development — and the implications this has for social-change strategies as well as for the role of universities and academics. I argue that a preoccupation with culture and (more recently) "hybridity" in the left camp and a preoccupation with limiting and controlling brown industry in the green camp have tended to divert our attention from the possibilities that exist for systematically redesigning and implementing sustainable and even *regenerative* agriculture, energy, and manufacturing systems. In contrast to older, quantitative definitions of wealth, we need to develop a *qualitative* concept of wealth — of wealth as *regeneration,* as the inner development of people, the (social) development of community, and the restoration of all living systems. In short, I will be suggesting that to address the current range of social, ecological, and economic problems that encircle us today, we may have to act on a fairly ecotopian agenda.

The Left and Productive Forces

It has been many decades since the intellectual left has had much to say about the *reconstructive possibilities* that might emerge out of capitalist economic and technological development. The left's relative silence on what Marxists refer to as the "productive forces" predates the rise of post-structuralism. An understandable aversion to mechanistic economism and historical determinism led many Western Marxists from the 1960s onwards to focus less on productive forces and more on the "relations of production" (Levine and Wright 1980). Such a manoeuvre to be sure provided more subtle and flexible understandings of social, cultural, and political change. However, it is increasingly clear that this "culture turn" — which continues today in various post-structuralist discussions of "hybridity" — has also had a cost. Notably, it has generated too little work that systematically explores the economic *alternatives* available to us.

By contrast, the first wave of popular "post-industrial" writing — by Daniel Bell (1976), Alvin Toffler (1972, 1980), and others — provided many insights into the emerging technological and economic potentials of a nascent post-industrial society; but this body of work also tended to rationalize many exploitative — and distinctly industrial — aspects of capitalism. Social radicals like Bookchin (1971, 1980), Touraine (1971), and Gorz (1971) did examine more critically the new developmental dynamics of post-industrialism — in particular, the implications for environmental and other social movements. But though these thinkers enjoyed some currency in late 1960s countercultures, they did not have much lasting impact among the academic or organizational left (Perrucci and Pilisuk 1968).

It could be argued that the real breakthroughs in radical post-industrial thought came from Radovan Richta and his colleagues (1969), Martin Sklar (1969), and Fred Block and Larry Hirshhorn (1979). From neo-Marxist standpoints, they highlighted the emerging centrality of culture in production and explored the implications of this for the relations among human, technological, and economic development. In particular, they outlined how culture challenged certain aspects of classical industrialism. They emphasized that industrial capitalism, which had always been based on "cog labour" and vast material throughput, would have to somehow manage new productive forces (NPFs) whose fundamental task would be to save resources and upgrade human creative capacities. These writers were, however, virtually ignored by an academic left that seemed jarringly oblivious to emerging grassroots social and ecological alternatives. In recent years the rise of new information technologies has spawned visionary new perspectives on wealth and change. A prime example of this is Benkler's (2006) *The Wealth of Networks*, which is written from a liberal-democratic perspective. Yet a more radical left analysis of these emergent forces — which range from open-source software to file sharing — has been largely non-existent, and this has left something of a vacuum regarding the political implications of these changes.

While the left's long-standing concerns about the distribution of economic and cultural power remain vitally important, a serious attempt to address issues of equality and ecology also demands a re-envisaging of different structures that could facilitate new forms of economic activity and a fundamental redefinition of wealth and production. I want to argue here that both the form and the content of economic life must be transformed, and that this will involve embedding social and ecological values into daily production and exchange as well as fostering new, decentralized forms of production and regulation. Such a transformation will be essential to creating authentic ecological economies; it may also constitute a realization of the traditional socialist project of reorienting the economy so that it directly serves human needs. In contrast to the mainstream environmentalist position, which views the relationship between human needs and ecosystem health as a trade-off, the radical green post-industrial perspective outlined here maintains that the regeneration of the natural world will depend on a more *direct targeting* of human needs. The post-scarcity economy and society I envision will use *less*, not more, material because it will centre production directly on human and environmental regeneration and because it will have the technical capacity to satisfy these needs with minimal resources. Later in this chapter I will return to this relationship between need and restructuring.

The New Productive Forces

Our current social and environmental crises are not simply a matter of the economic system's exploitative or destructive impacts; they are also a result of the suppression of vast and growing human potentials. The NPFs are basically forces of *people-production* — in contrast to the *thing-production* of past phases of industrial development—but they also make possible a new level of regenerative reintegration with non-human nature. A Marxist or feminist might view NPFs as aspects of the "reproduction of labour-power"—which they are—but we must appreciate that they go far beyond "simple labour-power" or industrial cog labour. They encompass a complex of creative, nurturing, giving, intuitive, artistic, healing, and co-operative capacities that have been systematically suppressed.

Capitalism is defined by property and power relationships that are generally antithetical to far-reaching human development, but competitive pressures have compelled the system at various points to at least selectively employ and cultivate NPFs. By the 1920s, industrialization had begun to move into the realm of culture, effecting significant changes in North America. Markets began to move beyond primary needs for large portions of the population, and production itself became increasingly comprised of cultural, organizational, and scientific work. The rise of mass-production industries placed questions of consumption (and, implicitly, the purpose of production) on the economic agenda in unprecedented ways, and new, more culturally defined social movements appeared.

Technology—understood as extensions of human senses and functions—also began to undergo a basic change. As emphasized by McLuhan (1962, 1964), technologies (such as heating) that had served mainly as extensions of our muscles and bodily functions now began to focus more on extending our minds and nervous systems through electronic media and information technology. In communications, the new cultural and informational forces first appeared under the old, linear, physical-production model—what Benkler calls the "industrial information economy." The exemplars of this model—telegraph, newspapers, radio, broadcast TV, and so on—required massive up-front investment and thus demanded equally vast markets in order to pay. Such a structure inevitably separated producers from largely passive consumers.

Over the past fifteen years we have begun to see the emergence of a new "networked information economy"—one that is quite different, and much more broadly distributed, than the industrial one. Today, powerful production hardware has been decentralized onto the desktops of ordinary citizens, making possible greater creative autonomy as new forms of mass collaboration and peer production begin to mushroom. Notwithstanding that capitalist globalization still seeks to expand the realm of old-line markets, equally powerful forms of demarketization are growing at the heart—not simply the periphery—of the information economy.

The NPFs therefore present some fundamental problems for capitalism. As Benkler has argued, old-line markets just don't work well to facilitate new forms of cultural production, which are based in non-material inputs and outputs. (Even earlier in industrialism's history, the role of universities testified to the inability of strictly market relationships to generate necessary knowledge production.) The expansion of knowledge and value production in the post-industrial context is creating a fundamental crisis of both the market economy and class society. Required, then, is a new relationship between resources and people, between material and non-material production, and between production and consumption.

Scarcity, Domination, and Quantitative Development

Economics, and an important dimension of ecology, are all about the relationship between people and resources. The elephant in the room whenever either social justice or ecological balance is discussed is material scarcity. Since a post-industrial economy is a post-material one, scarcity is a major consideration. It is a factor not simply because it affects whether human needs get satisfied, but because scarcity has been an essential tool of social control and labour discipline in civilization. For five thousand years, scarcity has been a prop of class power.

This fact is more or less acknowledged by the left, and it influences the left's emphasis on the fair distribution of wealth. However, the left tends to underestimate the importance of production—in content and structure—in maintaining this inequality. Especially now that cutting-edge production depends on the cultivation—not the suppression—of human creativity, the redistribution of material wealth and political power is impossible without transformations in what and how we are producing.

By contrast, many environmentalists fetishize production—especially through their emphasis on economic growth and overconsumption. By positing the "affluence" of whole nations as a fundamental cause of environmental destruction, they overlook the actual role played by inequality in this destruction, the role of scarcity in perpetuating this inequality, and the potential for qualitative or regenerative development. A politics based on limits rather than transformation (including democratic participation) risks reinforcing inequality, especially since the costs of crisis invariably fall most heavily on the powerless. Alienated personal consumption is definitely a problem, and individual responsibility can be an important starting point for activism; even so, such a perspective amounts to "blaming the victim" when it fails to take into account that a manipulative use of scarcity underpins the entire mass-consumption economy.

Marx saw the relationship between class society and *relative scarcity*. Civilization was made possible by a permanent economic surplus large enough to support classes but too small to provide abundance for all. A combination of elite control of scarce resources and elite monopoly of high culture defined power in civilization. For this reason, Marx saw capitalism as a contradictory system — that is, as a class society whose open-ended economic growth could eventually undermine class power by ending scarcity. Even at their most efficient, capitalist markets are driven by scarcity — which is not necessarily a bad thing at earlier stages of development, when most legitimate needs revolve around food, shelter, clothing, and basic infrastructure. But potential abundance changes everything, in part by undercutting traditional market drivers. This is why the Great Depression can be understood as a reaction to the threat of abundance — as a spontaneous system shutdown following the unprecedented productivity explosion of the Roaring Twenties. While a number of factors contributed to the Great Crash of 1929, the failure of recovery was due primarily to a new structural crisis of overproduction: business was simply *not confident* that there would be sufficient "effective demand" for the vast productive capacity that had been developed. The emerging stage of economic evolution called for a more needs-focused mode of qualitative development, and this was something that capitalism was not, and may not be, capable of developing.

Marx recognized that material abundance could potentially sabotage class society, but he probably did not anticipate the possibility that growth could be redirected into waste in order to reinforce scarcity and class. This is essentially the story of postwar (or "Fordist") North American capitalism, which, contrary to the popular notion of creating "the Affluent Society" (Galbraith 1958), actually served to artificially recreate scarcity through the production of waste — *the Effluent Society*. A progressive solution to the new crisis of effective demand would have entailed, among other things, greater redistribution of income, thereby increasing worker purchasing power — one of the key proposals of left New Dealers. Real abundance, however, would also have entailed a more *qualitative* redefinition of wealth — a call echoed at that time by labour and social-movement advocacy of work-time reduction, universal free education and health care, and intelligently planned communities. This last would have been an authentic step toward real abundance, but alas, such options were soundly defeated by around 1948, as the Levittown model of suburban sprawl and the permanent war economy became the twin pillars of North American capitalism.

Many activists today wax nostalgic for postwar Fordism — understandably, since the system seemed to allow worker and citizen gains, which were the cumulative outcome of union and social-movement struggles over several decades. These gains included the legitimization of collective bargaining in key industries, the creation of social safety nets, and so on. But the new "waste economy" consistently took with one hand what it gave with the other. Kunstler (1994) calls the postwar suburban boom "the greatest misallocation of resources in human history." The unnecessary waste on which the auto/oil/suburb complex was founded helped create a twenty-five-year boom; that boom's costs would eventually come due beginning in the 1970s. Meanwhile, all of this waste and alienated consumption was financed by planned inflation as well as by new forms of "debt money" that kept workers tied to a debilitating work-and-spend cycle even in the best of times. Fordism's "paper economy" (Bazelon 1963) laid the foundations for the post-Fordist "casino economy" of empty financialization that collapsed in the second half of 2008.

On the surface it may seem unrealistic to suggest that humanity can or should move beyond quantitative development. But the material scarcity of today is quite different from that of even one hundred years ago. Today it is artificially created and maintained. Going beyond scarcity now involves going beyond accumulation and establishing new forms of qualitative development everywhere. Especially in the most poverty-stricken areas of the world, "income growth" can be a poor measure of economic well-being: it often disguises the erosion of traditional livelihoods and appropriation of regional resources by global corporations. Indicators of *how well needs are met* are far more reflective of real wealth and efficiency. Today's "disaster capitalism" (Klein 2008) is grounded in a psychology and politics of scarcity and fear — even in the best of times. A green economy needs to be grounded in a culture of abundance and plenitude. This is no contradiction of the equally important conserver mentality: real abundance and fulfillment are qualitative, not quantitative. People are far more willing to accept material limits when they have access to many opportunities for community, real security, and personal fulfillment.

Decentralization and Post-Industrialism

The left has been slow to recognize the role of waste in holding back real development. So, too, has it underestimated the decentralist character of post-industrialism.

The industrializing economy during its nineteenth-century heyday was chaos, a runaway freight train. Despite its apparently decentralized character, it was capital-intensive, and its developmental tendency was toward centralization. One part of this was the technological centralization encouraged by industrialism's concentrated energy sources, especially fossil fuels. Measures to mitigate the chaos or to encourage a more humane and egalitarian industrialization would have had to be even more strongly centralist, as were most socialist movements. Efforts to control the economy were further complicated by the structural chasms separating economics, politics, and culture in early capitalism. For the working class and various popular movements, real change required a *dual strategy*, one part economic, the other political. The labour movement could focus on production, but it also needed a political arm — be it a socialist, labour, or democratic. The basic strategies of the labour and social movements had to be more oppositional because of the importance of the state to possible alternatives. In any case, "alternatives" for the nineteenth-century working class likely meant a more equitable sharing of existing wealth, along with better conditions of work. Few had any quibbles with the material character of wealth.

Today the most conscious elements of all the social movements are disputing capitalism's very definition of wealth. They are contesting the *nature* of production, not just its distribution. Equally important, they are in a position to launch economic alternatives themselves, without prior control over, or influence on, the state. Three key interrelated dimensions of authentic post-industrialism have made this decentralization possible.

The first of these is the pressure toward democratization, human scale, and grassroots participation. Notwithstanding the power of corporate globalization and religious fundamentalism, democracy is an inextricable element of the NPFs; throughout the world, community is being increasingly recognized as the nexus of real development. This means not simply a growing role for civil society, but a fundamental re-embedding of the economy in civil society and community. A North American example is the movement for "local living economies," which in the present day is growing even faster than the movement for corporate sustainability.

Second are pressures toward efficiency and harmony with nature. Industrialism's fossil-fuel-based centralization supported a linear, extraction-to-disposal resource cycle. By contrast, emergent sustainable agriculture and food systems, soft-energy infrastructures, ecological waste-resource systems, holistic health systems, and sustainable urban design — and more — tend to involve both greater decentralization and circular resource flows. Even ecomanufacturing — which is based in reuse, services, and regionally benign materials — demands proximity and increasingly localized closed loops (Stahel 1994). This involves not simply more efficient use of energy and resources, but a restructuring of economic processes so that they move within, or imitate, natural systems — what has been called "economic biomimicry." It also entails a more direct focus on *end-use* or human need, as well as a process of "back-casting" to find the most elegant and efficient ways of meeting that need. As soft-energy guru Amory Lovins (1977) put it, we want "hot showers and cold beer," not necessarily fossil fuels and power plants. The latter are just means to the end. Industrial ecologists have called this the "ecoservice economy," which is geared to meeting needs for nutrition, access, illumination, entertainment, and so on, and which has the potential to conserve resources in fundamental ways. This is why advocates argue that harmony with non-human nature is not only not counter to (or a trade-off with) human interests, but in fact *depends on* a more direct targeting of real human needs.

The third main dimension of post-industrial decentralization is rooted in the character of knowledge-based production. Ecological economist Herman Daly (2000) points out that real economic efficiency today necessitates greater restrictions on the flow of material goods and physical capital (to minimize transportation costs, to optimize the use of local resources, etc.), and fewer restrictions on the flow of knowledge. As he says, "trade recipes, not cookies." According to Daly, this common sense runs directly counter to ongoing capitalist globalization, which, through free trade and intellectual property rights, is working directly against economic intelligence. While much of the left tends to focus on the destructiveness of free trade, it may be that intellectual property law is an even more insidious form of repression and waste. Analysts of the new knowledge economy, such as Benkler, point out that the decentralization of information hardware is creating new forms of mass collaboration and peer production outside the market economy. The most obvious examples of this are Wikipedia, open-source software, and all kinds of file sharing. Capitalism's current attempts to suppress this sort of swapping (via a new wave of intellectual property law and campaigns against "piracy") amount, in the words of Lawrence Lessig (2004), to a "war on creativity" — and thus a new level of waste of human productive capacity. In the era of physical production, copyright and patent law served to support innovation; today it tends to suppress innovation in the financial interests of the corporate elite. As Brand said, "information wants to be free" and to costlessly multiply itself (DeKerckhove 1995). In this sense the new information commons seems more consistent with "gift economy" relationships of the sort advocated by feminists like Genevieve Vaughan (1997) than with conventional, scarcity-based market exchange. In any case, the demarketization that is taking place at the cutting edge of knowledge-based production is putting the lie to the corporate stance that the commodification of life is the essence of progress.

In recent times, economic crisis has highlighted the irrationality of capital's relations with resources and information. The post-Fordist era, beginning with Reagan and Thatcher, introduced new and even more empty forms of waste production. It channelled new information technology into financialization and into a global casino economy based in fictitious capital, illusory debt money, and speculation. Like the planned inflation and paper economy of the Fordist period, it has essentially been a means of "reredistributing" wealth back to those with money-making powers, and critics have long seen it as a kind of Ponzi scheme, a house of cards that would eventually topple (and that did topple in 2009). By the same token, globalization's irrationally overextended loops of production and consumption — as manifest in Wal-Mart and the China trade — seem threatened by the spectre of "peak oil." Localization, once seen as counterculture utopianism, is now viewed as a legitimate strategy by growing numbers of economic planners, policy-makers, and analysts. Business networks such as the Business Alliance for Local Living Economies (BALLE) have begun to make small but tangible and influential impacts on local economies, and this has encouraged a new wave of interest in the "local multiplier" effects of investment — that is, in comparing locally owned with global companies (Shuman 2006). That said, the creation of self-reliant local economies is certainly not a *sufficient* condition for socially just ecological development, or for the full expression of the NPFs.

Distributed Regulation and Mindful Markets

Localization and the re-embedding of the economy in civil society imply basic changes in the nature of regulation and of the state. A different style of regulation is also suggested by the fact that so much green production could be *regenerative*, and not simply a toxic destructive juggernaut that nature and community must be protected from. Moreover, the very complexity of the contemporary economy (especially one based in quality) defies a concentration of regulatory power in the state. Complex systems tend to require greater measures of internal self-regulation based in sophisticated feedback loops. No less than soft energy systems or sustainable food production, green regulation must be *distributed*, as well as "environmental" in the sense of being structured into a range of incentives and disincentives embedded in everyday economic choices and relationships.

This is definitely not an argument for the voluntarism typically advocated by corporate critics of "command and control" regulation — a transparent rationalization for less corporate accountability. If anything, post-industrial planning and regulation will have to be *more* thoroughgoing and conscious, but this will have to take place on a multitude of levels, some of them not usually associated with the state. What the state should be trying to accomplish and express is a new regenerative economic balance of competition and co-operation. Unhealthy forms of competition are blocking alternatives — for example, eco-industrial networks that require substantial sharing of information and decision making. Such competition is also behind the massive waste that results from the needless incompatibility of all kinds of electronic equipment. By the same token, healthy competition to increase quality is, today, often stymied by monopoly corporate and financial interests. Changing the competitive/ co-operative balance is what many of the new enterprise networks, such as the Business Alliance for Local Economic Development in Toronto, are attempting to achieve (http://www.livingeconomies.org). Co-operation between ostensible competitors is creating ripple effects that are jump-starting entire regenerative industries such as locally sustainable food systems. These industries are quite different from old-line industry associations. Governments have many crucial tools available to reorient incentive structures; but their success will depend on being tuned to the pulse of society, including to the new enterprise networks and all other stakeholders. This "regulatory pluralism" — characterized by openness and flexibility — would transform the very drivers of economic life, including the "DNA" of enterprises, which increasingly are being programmed for regenerative activity.

Local Finance, Green Jobs, and the Green-Collar Economy

One powerful force that must explicitly act as a regulatory mode is finance. There will need to be preferential lending for activities and enterprises that correspond to well-defined green community plans. There is something to be said for "green jobs." In Jones's (2008) view, we have entered the "investment" phase of environmentalism, having passed through two previous phases: "regulation" (1970s) and "conservation" (1800s). By this, Jones means not that we have moved beyond regulation or conservation, but that our primary need is for fundamental economic restructuring. This is one reason why the concern of some critics (e.g., Nordhaus and Shellenberger 2007) — that the environmental movement has not matched its regulatory achievements of the 1970s — is not really that relevant. At a different level, the environmental movement has made substantial gains in developing ecological alternatives in basic sectors. Today, however, more comprehensive means are required to channel resources in order to "mainstream" these alternatives.

Finance itself would be only one part of the new incentive structure of a green economy. For a government, taxes, procurement, and infrastructure can all be important means of encouraging good things. Also, as discussed earlier, the very scale of the economy can encourage greater accountability — especially when local and regional governments refrain from giveaways to non-local corporations and chains. But other kinds of rules and relationships are also necessary, especially those that institutionalize stakeholder power. Mainstream corporate social responsibility (CSR) pays much lip service to stakeholder consultation; but processes will be needed that guarantee that all stakeholders will have some degree of decision making over activities that impact their lives. Marjorie Kelly (2002), the former editor of *Business Ethics*, wrote that "economic democracy is the next stage of corporate social responsibility." In a green economy, participation can rarely be separated from other dimensions.

With regard to CSR and green business, there is a growing scholarly literature on new forms of "non-state governance" and regulatory pluralism. These forms are expressed in developments such as food, building, and wood certification. Many of these authors, however, underestimate the value revolution that the expanded tool box of regulatory instruments must ultimately accomplish. Over the long haul, the goal must be to displace profit and accumulation as primary market drivers. Cashore (2004), for example, views green third-party certification systems as a form of "non-state *market-driven* governance," without fully appreciating that the best of these systems manifest a new form of *market* as much as a new form of *governance*. When these certification systems are truly doing their job, the market drivers they create are primarily social and ecological, not financial. Profit is by no means incidental or unimportant, but is now a *means* to realize qualitative value rather than the end itself. This is the ultimate meaning of "market transformation" — what Korten (1999) refers to as the creation of "mindful markets." This is especially clear in one of the most radical developments of the certification movement: that of the certification of business-governance systems. "B Corporation" is an example of this kind of certification, which requires certified companies to embed social and environmental commitments into their governing documents. This entails taking on substantial new liabilities; but at the same time, it provides the companies with markets and other network services. Like the corporate-charter movement, B Corporation attempts to deal with the structural "pathology" (illustrated in the documentary *The Corporation;* Achbar, Abbott, and Baken 2003) of the publicly traded limited-liability company chained to producer profit for shareholders. It does this, however, in a positive way — by emphasizing change in a firm's "DNA" rather than the imposition external restrictions on companies.

One of the fundamental challenges of the new green regulation is to connect responsibility with regeneration, which means redefining ownership to support stewardship. Abstract questions of public versus private can be distracting holdovers from the industrial era. From a green economic-design perspective, the question is how ownership can be shaped to support stewardship. Extended producer responsibility (EPR) is perhaps the most important policy principle of the "life-cycle approach," next to the precautionary principle. In its classic application to, for example, consumer durables, EPR means centralized ownership, with producers on the financial hook for their products' environmental impacts over their entire life cycle. Customers purchase the service but lease the product. Typically, this encourages producers to design for disassembly and to detoxify their products, in order to minimize waste and life-cycle costs. This shift normally increases the labour component and decreases the resource component of production; it also encourages regionalism and upgrades stakeholder relationships. But EPR is a principle that has to be applied differently to different situations. There are many cases where decentralized ownership—for example, small holding—is most conducive to stewardship. The state has a major role to play in determining the appropriate balance of kinds of ownership. But on the whole, in a green system of distributed regulation, government is free to act less as a policeman and more as a coordinator.

Strategic Priorities for Action: For the Economy and Universities
Despite the depth and scope of the crises we face, making fundamental social change today demands an increasingly positive mindset, for two interrelated reasons. First, there is the appreciation that NPFs are grounded in all-round human development and that they permit and require growing levels of decentralization. Second, there is the recognition of the two key dimensions of a green or post-industrial economy: (1) it must be primarily a *service economy*, organized for the direct targeting and satisfaction of human needs; and (2) it has to be a more *circular, closed-loop system*, one that is deeply integrated with nature—a sailboat in the winds of natural processes.

These new realities have radically altered social-change strategies compared to those of the industrial era, which had to be more oppositional. Marx basically saw the working-class revolution as happening before the appearance of NPFs; indeed, it was the successful revolution's task to ensure that NPFs were developed and implemented. But NPFs began to emerge *before* the revolution, fundamentally changing revolutionary strategies. In Marx's time, the working class had to accomplish the revolution as the first step toward creating the Good Society and the New Human, whereas today the order is reversed: the working class must create the Good Society and the New Human *in order to* make the revolution. Social and ecological alternatives become the centrepiece of movement strategy, with necessary oppositional activity geared to support this focus. "Seizing the means of production" is still necessary, but today the means of production are ourselves and our full creative capacities.

Despite the intellectual left's slowness to acknowledge emerging potentials, it is well placed to make substantial contributions to the tasks at hand, especially in universities. Because of the knowledge intensity of post-industrial production, universities have key roles to play in the production, evaluation, training, and regulation of green economies. Because of their size, coupled with the need to increase localization, universities also have a major role to play in consumption, or market creation, within local economies.

The development of new measures of qualitative value — and their economic applications — is crucial to post-industrial development. These measures include mass-balance accounts, eco-footprints, product life-cycle analysis or LCA, firm eco-accounting, genuine-progress indicators, local economic multipliers, and sustainable-community indicators. Universities should be an important base not just for the study of emergent third-party certification systems (as suggested by Cashore), but for their creation. Another essential that universities could contribute to is comprehensive directories and databases of regenerative products and services. These virtual marketplaces could be powerful tools in what will likely emerge over the next decade as one of the most significant strategic initiatives of green economic development: Local First (buy local) campaigns, organized by new community business networks and their diverse allies (Shuman 2006).

Left intellectuals, embodying a heritage of critical political-economic thinking, can also provide "big picture" thinking. This can be vital at a time when Wal-Mart–style efficiency measures are hailed as a sustainability revolution. The "service and needs" focus of a green economy is essentially post-industrial socialism. This is not socialism as industrial-state formation, but as a philosophy of the priority of human needs over blind market forces. Such a tradition can be invaluable to understanding and supporting qualitative development today. Critical thinking about markets can be vital today, when non-market and "mindful market" production and exchange is exploding. Many key forms of ecological production — energy retrofit, rooftop and community gardens, preventive health care, and so on — take place in or around the home, in the informal economy. In today's economy, these things are viewed as forms of consumption and are economically penalized; instead we have to find ways to properly remunerate or support these essential activities. At the same time, we have to find ways to support the emergent electronic commons in a way that does not burden it with restrictive market relationships.

One of the great priorities for post-industrial development is the achievement of new forms of economic security. The earlier discussion of the role of scarcity in suppressing progress only hints at the actual role played by fear and insecurity in maintaining waste, exploitation, and domination. In tough times, people will do nasty things for money. Even when times are good, many people who have no interest in big money or power are driven to compete and accumulate simply because they cannot trust society to take care of them when they are old or ill. They cannot trust that their children will get a decent education.

Possible solutions include basic- or citizen-income schemes (i.e., a guaranteed annual income). By fully meeting people's basic needs, these programs might unleash a tide of creative activity hitherto held back by fear and uncertainty. Carbon and other kinds of green taxation are appropriate sources of funding for basic-income schemes, since one goal of such schemes is a fundamental shift in investment from resources to people. As with many forms of green production, the approach is to employ savings as a source of finance. Community currencies are another fertile ground for experimenting with support for healthy forms of non-market production. In our electronic environment we can expect to see imaginative and sophisticated new forms of economic exchange, all geared to reclaiming the information revolution from contemporary capitalism's casino economy.

In a complex society, many forms of public goods, new and old, are *not* financed most appropriately by user-pay, market prices, or wages. Benefits are spread so widely that it makes sense for governments or communities to provide for them, financed by taxes or whatever. As the music industry has already found out, the electronic commons is displacing many unnecessary market relationships, and we should be making more conscious choices about how we might facilitate this in constructive ways. In particular, we need to find out what kinds of economic security might unleash *giving*—and not simply reciprocity—as a major force in economic life.

The biggest current impediment to universities tapping their potential as incubators of qualitative wealth is probably their attitudes toward localization and globalization. Qualitative wealth is based strongly on place, but the function of the industrial university has been increasingly to service big corporate enterprises, along with the state infrastructures that support them. In the past several years there has been heated debate over the future of higher education: Should it be "job training" or "liberal education"? Many people on the left are justifiably concerned about the subordination of education to corporate needs, so they argue against a focus on job training. But a greater emphasis on practical job skills is not in itself a bad idea, so long as those skills are geared toward meeting real community needs. Such work goes hand in hand with the best kind of liberal education, since all-round human development is actually the most direct route to post-industrial productivity. One cannot make this case for corporate–global development, where universities are asked to produce intellectual labourers for work that is largely destructive of community and environment.

In the corporate–global university, social-change–oriented academics have been increasingly relegated to the realm of analysis and critique, not actual production. How many North American university faculty are involved in creating sustainable food systems, designing soft-energy systems, supporting green building, or initiating eco-industrial development? How many students, undergrad or grad, are concretely contributing to community development in their studies and research? The university itself remains unaware of its important role in the local economy. Simply by basing its purchasing decisions on social and environmental values, a university could radically affect the ability of local economies to create regenerative work, foster dynamic local enterprises, and plug debilitating leaks in the regional economy. Fortunately, there is a rapidly growing "university sustainability" movement, but even here many of these initiatives fail to acknowledge their impacts, existing and potential, on the economies of the communities in which they are situated.

Conclusion

Those who belong to the intellectual left have much to offer green and post-industrial transformation. But first they will have to acknowledge that transformation, including its direction and the role that knowledge will play in it. We are talking about far more than environmental protection here. A retreat from broad-based vision is not an option. What we need now is a positive embracing of qualitative change that can be a service to all progressive social movements and to higher education.

Works Cited

Achbar, M., J. Abbott, and J. Baken. 2003. *The Corporation.* Big Picture Media Corporation, Canada. 145 min.

Bazelon, D. 1963. *The Paper Economy.* New York: Vintage.

Bell, D. 1976. *The Coming of Post-Industrial Society: A Venture in Social Forecasting.* New York: Basic.

Benkler, Y. 2006. *The Wealth of Networks: How Social Production Transforms Markets and Freedom.* New Haven, CT: Yale University Press.

Block, F., and L. Hirschhorn. 1979. "The New Productive Forces and the Contradictions of Contemporary Capitalism." *Theory and Society* 7: 363–90.

Bookchin, M. 1980. *Towards an Ecological Society.* Montreal: Black Rose.

———. 1971. *Post-Scarcity Anarchism.* Berkeley, CA: Ramparts.

Cashore, B.W. 2004. *Governing Through Markets: Forest Certification and the Emergence of Non-State Authority.* New Haven, CT: Yale University Press.

Daly, H. 2000. "Globalization and Its Discontents." Paper based on a discussion at the Aspen Institute's 50th Anniversary Conference, "Globalization and the Human Condition," August 20, 2000, Aspen, CO.

DeKerckhove, D. 1995. *The Skin of Culture: Investigating the New Electronic Reality.* Toronto: Somerville House.

Galbraith, J.K. 1958. *The Affluent Society.* Boston: Houghton Mifflin.

Gorz, A. 1971. *Strategy for Labor: A Radical Proposal.* Boston: Beacon.

Jones, V. 2008. *The Green-Collar Economy: How One Solution Can Fix Our Two Biggest Problems.* New York: Harper One.

Kelly, M. 2002. "It's a heckuva time to be dropping business ethics courses." *Business Ethics* 16, nos. 5 and 6: 17–18.

Klein, N. 2008. *The Shock Doctrine: The Rise of Disaster Capitalism.* New York: Metropolitan.

Korten, D. 1999. *The Post-Corporate World: Life after Capitalism.* San Francisco: Berrett-Koehler.

Kunstler, J.H. 1994. *Geography of Nowhere: The Rise and Decline of America's Man-Made Landscape.* New York: Touchstone.

Lessig, L. 2004. *Free Culture: The Nature and Future of Creativity.* New York: Penguin.

Levine, A., and E.O. Wright. 1980. "Rationality and Class Struggle." *New Left Review* 123: 47–68.

Lovins, A. 1977. *Soft Energy Paths.* New York: Harper Colophon.

McLuhan, M. 1964. *Understanding Media: The Extensions of Man.* New York: Mentor.

———. 1962. *The Gutenberg Galaxy.* Toronto: University of Toronto Press.

Nordhaus, T., and M. Shellenberger. 2007. *Breakthrough: From the Death of Environmentalism to the Politics of Possibility.* New York. Houghton Mifflin.

Perrucci, R., and M. Pilisuk. 1968. *The Triple Revolution: Social Problems in Depth.* Boston: Little, Brown.

Richta, R., et al. 1969. *Civilization at the Crossroads: Social and Human Implications of the Scientific and Technological Revolution.* Prague: International Arts and Sciences Press.

Shuman, M.H. 2006. *The Small-Mart Revolution: How Local Businesses Are Beating the Global Competition.* San Francisco: Berrett-Koehler.

Sklar, M.J. (1969) "On the Proletarian Revolution and the End of Political-Economic Society." *Radical America* 3, no. 3 (1969): 1–41.

Stahel, W.R. 1994. "The Utilization-Focused Service Economy: Resource Efficiency and Product-Life Extension." In *The Greening of Industrial Ecosystems,* ed. B.R. Allenby and D.J. Richards. Washington: National Academy Press. 178–90.

Toffler, A. 1980. *The Third Wave.* New York: Bantam/William Morrow.

———. 1972. *Future Shock.* New York: Bantam.

Touraine, A. 1971. *The Post-Industrial Society: Tomorrow's Social History: Classes, Conflicts, and Culture in the Programmed Society.* New York: Random House.

Vaughan, G. 1997. *For-Giving: A Feminist Critique of Exchange.* Austin, TX: Plain View.

Sklar, M. J. (1992). "On the Proletarian Revolution and the End of Political-Economic Society." Radical America 3, no. 3 (1969): 1–41.

[illegible] (19[illegible]). The Utilization [illegible] and Its Extension. In The Greening of [illegible] [illegible] Nationalism [illegible] [illegible].

[illegible]

[illegible]

Contributors

Aidan Davison is a lecturer in human geography and environmental studies at the University of Tasmania. His interdisciplinary research interests arise at intersections of socio-cultural themes of nature, technology, and sustainability. The author of *Technology and the Contested Meanings of Sustainability* (Albany, NY: SUNY Press, 2001), he has published many articles and book chapters on topics such as public perceptions of biotechnology, Australian environmentalism, and education for sustainability.

Simon Guy is a professor of architecture at the University of Manchester. His research aims to critically understand the co-evolution of design and development strategies and socio-economic processes shaping cities. His publications include (with S. Moore) *Sustainable Architectures: Cultures and Natures in Europe and North America* (Oxford: Spon, 2005) and (with Elizabeth Shove) *A Sociology of Energy, Buildings, and the Environment: Constructing Knowledge* (London: Routledge, 2000).

Steve Hinchliffe is a reader in environmental geography and director of research for geography at the Open University. He works on the geographies of nature, non-humans, and environments. He is author and editor of numerous books and articles on issues ranging from risk and food to biosecurity, urban ecologies, and nature conservation. His research focuses on the "making of things in practices" and draws together insights from science and technology studies (STS) and geography. His publications include *Geographies of Nature: Societies, Environments, Ecologies* (London: Sage, 2007); and (with Kathryn Woodward) *The Natural and the Social: Change, Risk and Uncertainty*, second edition (Oxford: Oxford University Press, 2004).

Fletcher Linder is an associate professor of anthropology at James Madison University. He has studied and published across a variety of topics, including sports and aesthetics, illness experience and care, interpersonal violence, and environmental politics. He has conducted ethnographic, epidemiological, urban-landscape, and community-based intervention research in such areas as the American South, California, Canada, and Australia. He is presently completing a monograph titled "Waiting for Arnold: Image, Body Discipline, and Late Capitalism."

Timothy W. Luke is University Distinguished Professor of Political Science at Virginia Polytechnic Institute and State University in Blacksburg, Virginia. He also is the Program Chair for Government and International Affairs in the School of Public and International Affairs, and founding Director of the Alliance for Social, Political, Ethical, and Social Theory (ASPECT) in the College of Liberal Arts and Human Sciences at Virginia Tech. His publications include *Capitalism, Democracy and Ecology: Departing from Marx* (Champaign: University of Illinois Press, 1999); *The Politics of Cyber Space* (co-edited with Chris Toulouse — New York: Routledge, 1998); and *Eco Critique: Contesting the Politics of Nature, Economy and Culture* (Minneapolis: University of Minnesota Press, 1997). The author of more than 150 journal articles and edited book chapters, he writes extensively on the politics of museums as well environmental politics, international affairs, and social theory.

Mike Michael is a professor of sociology of science and technology, and director of the Centre for the Study of Invention and Social Process, in the sociology department, Goldsmiths, University of London. His research is concerned with a number of areas, notably the public understanding of science; the sociology of mundane technologies; the sociology of biomedical innovation; the sociology of everyday life; animals and society; and materiality and sociality. He is the author of *Technoscience and Everyday Life: The Complex Simplicities of the Mundane* (Bristol: Open University Press, 2006); *Science, Social Theory, and Public Knowledge* (with Alan Irwin — Bristol: Open University Press, 2003); *Reconnecting Culture, Technology, and Nature: From Society to Heterogeneity* (London: Routledge, 2002); and *Constructing Identities: The Social, the Nonhuman, and Change* (London: Sage, 1996).

Brian Milani is an associate of the Transformative Learning Centre and coordinator of the Business and Environment Program at York University's Faculty of Environmental Studies. He is author of *Designing the*

Green Economy (Lanham: Rowman and Littlefield, 2000) and a member of the Coalition for a Green Economy. His focus for more than two decades has been on creating grassroots ecological alternatives through community development, construction, education, and general trouble making. He was co-founder of Green City Construction and is the director of Toronto's long-running course on green economic alternatives, "The Green Economy at the Labour Education Centre," featuring Toronto's cutting-edge eco-innovators. He has also been involved with green labour activities at the Labour Council of Toronto and Carpenters Local 27.

Peter Oosterveer is a senior lecturer in environmental policy in the Department of Social Sciences at Wageningen University. He has published extensively on globalization and the sustainability of food production and consumption; the labelling and certification of food; environmental policy and management in Africa; and social theory and "a sociology of flows."

Erik Swyngedouw is a professor of geography at the University of Manchester's School of Environment and Development. From the late 1980s until 2006 he taught at Oxford University, latterly as Professor of Geography, and was a Fellow of St. Peter's College. His research focuses on political-economic analysis of contemporary capitalism. He has produced several major works on economic globalization, regional development, finance, and urbanization. Recently his interests have turned to political-ecological themes and the transformation of nature, notably water issues, in Ecuador, Spain, Britain, and elsewhere in Europe. His publications include *Globalizations* (Philadelphia: Temple University Press, 2004); *Social Power and the Urbanization of Water—Flows of Power* (Oxford: Oxford University Press, 2004); and (with F. Moulaert and A. Rodriguez, eds.), *The Globalized City: Economic Restructuring and Social Polarization in European Cities* (Oxford: Oxford University Press, 2003).

Julie Sze is associate professor of American studies at the University of California, Davis, as well as the founding director of the Environmental Justice project for UC Davis's John Muir Institute for the Environment. Sze's book, *Noxious New York: The Racial Politics of Urban Health and Environmental Justice,* won the 2008 John Hope Franklin Publication Prize, awarded annually to the best published book in American studies. Sze's research investigates environmental justice and environmental inequality; culture and environment; race, gender, and power; and community health and activism. She has published on a wide range of topics such as energy and air polution activism; toxicity; the cultural politics of the Hummer, and on environmental justice novels and cultural production.

Sarah Whatmore is a professor of geography and director of the International Graduate School at the Oxford University Centre for the Environment/School of Geography. Her research focuses on relations between people and the material world, particularly the living world, and the spatial habits of thought that inform the ways in which these relations are imagined and practised in the conduct of science, governance, and everyday life. She has published widely on the theoretical and political implications of these questions in the fields of agriculture and food; land rights and land-use planning; and biodiversity and biotechnology. These themes are brought together in her most recent books: *Hybrid Geographies: Natures Cultures Spaces* (London: Sage, 2002); *Using Social Theory: Thinking Through Research* (co-edited with Michael Pryke and Gillian Rose — London: Sage, 2003); and *Cultural Geography: Critical Concepts*, 2 vols. (co-edited with Nigel Thrift — London: Routledge, 2004). She received the Cuthbert Peek award from the RGS/IBG in 2003 for "innovative contributions to the understanding of nature–society relations."

Damian F. White is an assistant professor of sociology in the Department of History, Philosophy, and Social Science at the Rhode Island School of Design (RISD). Prior to coming to RISD, he was an assistant professor of sociology at James Madison University; a post-doctoral research fellow in the Department of Innovation Studies, University of East London, working on the European Union project "Optimising the Public Understanding of Science"; and a lecturer in sociology at Goldsmiths College, University of London. He has published articles on the historical relations between human societies and nature; ecotechnology and the "green industrial revolution"; the "production of nature" debate; anti-environmentalism; and the libertarian and anti-authoritarian traditions of the political left. He is the author of *Murray Bookchin: A Critical Appraisal* (London: Pluto Press, 2008) and, with Alan Rudy and Brian J. Gareau, the author of *The Environment, Nature and Social Theory* (London: Palgrave Macmillan, forthcoming).

Chris Wilbert is a senior lecturer in tourism and geography in AIBS at Anglia Ruskin University, England. He has published articles on animal geographies, political ecology, climate change and leisure/tourism. He is co-editor (with Chris Philo) of *Animal Spaces, Beastly Places: New Geographies of Human–Animal Relations* (London and New York: Routledge, 2000) and *Killing Animals* (Champaign: University of Illinois Press, 2006) with the Animal Studies Group. Recent articles have focused on the politics of avian flu in southeast Asia in *Focas: Forum on Contemporary Art & Society* 6 (Special Issue on Regional Animalities, 2007), and on crime scene tourism (with Rikke Hansen) in the book *Strange Spaces: Explorations into Mediated Obscurity* edited by André Jansson and Amanda Lagerkvist (Aldershot: Ashgate, 2009).

Index

Environmental Humanities Series

Environmental thought pursues with renewed urgency the grand concerns of the humanities: who we think we are, how we relate to others, and how we live in the world. Scholarship in the environmental humanities explores these questions by crossing the lines that separate human from animal, social from material, and objects and bodies from techno-ecological networks. Humanistic accounts of political representation and ethical recognition are re-examined in consideration of other species. Social identities are studied in relation to conceptions of the natural, the animal, the bodily, place, space, landscape, risk, and technology, and in relation to the material distribution and contestation of environmental hazards and pleasures.

The Environmental Humanities Series features research that adopts and adapts the methods of the humanities to clarify the cultural meanings associated with environmental debate. The scope of the series is broad. Film, literature, television, Web-based media, visual art, and physical landscape—all are crucial sites for exploring how ecological relationships and identities are lived and imagined. The Environmental Humanities Series publishes scholarly monographs and essay collections in environmental cultural studies, including popular culture, film, media, and visual cultures; environmental literary criticism; cultural geography; environmental philosophy, ethics, and religious studies; and other cross-disciplinary research that probes what it means to be human, animal, and technological in an ecological world.

Gathering research and writing in environmental philosophy, ethics, cultural studies, and literature under a single umbrella, the series aims to make visible the contributions of humanities research to environmental studies, and to foster discussion that challenges and reconceptualizes the humanities.

SERIES EDITOR:
Cheryl Lousley, English and Film Studies, Wilfrid Laurier University

EDITORIAL COMMITTEE:
Adrian J. Ivakhiv, Environmental Studies, University of Vermont
Catriona Mortimer-Sandilands, Tier 1 CRC in Sustainability and Culture, Environmental Studies, York University
Susie O'Brien, English and Cultural Studies, McMaster University
Laurie Ricou, English, University of British Columbia
Rob Shields, Henry Marshall Tory Chair and Professor, Department of Sociology, University of Alberta

FOR MORE INFORMATION, CONTACT:
Lisa Quinn
Acquisitions Editor
Wilfrid Laurier University Press
75 University Avenue West
Waterloo, ON N2L 3C5
(519) 884-0710 ext. 2843
Email: quinn@press.wlu.ca

**Titles in the Environmental Humanities Series
from Wilfrid Laurier University Press**

Animal Subjects: An Ethical Reader in a Posthuman World, edited by Jodey Castricano / 2008 / x + 314 pp. / ISBN 978-88920-512-3

Open Wide a Wilderness: Canadian Nature Poems, edited by Nancy Holmes / Introduction by Don McKay / 2009 / xxviii + 516 pp. / ISBN 978-1-55458-033-0

Technonatures: Environments, Technologies, Spaces, and Places in the Twenty-first Century, edited by Damian F. White and Chris Wilbert / 2009 / xii + 270 pp. / ISBN 978-1-55458-150-4